Beginner's Guide to
Reading
Schematics

Beginner's Guide to
Reading
Schematics

Fourth Edition

Stan Gibilisco

New York   Chicago   San Francisco   Athens   London
Madrid   Mexico City   Milan   New Delhi
Singapore   Sydney   Toronto

Beginner's Guide to Reading Schematics, Fourth Edition

9   LCR   23   22

ISBN   978-1-260-03110-2
MHID   1-260-03110-1

Sponsoring Editor	**Acquisitions Coordinator**	**Indexer**
Michael McCabe	Elizabeth Houde	Claire Splan
Editing Supervisor	**Project Manager**	**Art Director, Cover**
Donna M. Martone	Patricia Wallenburg	Jeff Weeks
Production Supervisor	**Proofreader**	**Composition**
Pamela A. Pelton	Claire Splan	TypeWriting

About the Author

Stan Gibilisco, an electronics engineer, mathematician, and radio hobbyist, has authored numerous titles for the McGraw-Hill *Demystified* and *Know-It-All* series, along with dozens of other technical books and magazine articles. His work appears in several languages in countries throughout the world. Stan has been an active amateur radio operator since 1966. His currently holds the call sign W1GV.

In Memory of Jack

Contents

Introduction

Have you "caught the electronics bug" and then balked at the sight of dia-grams with arcane symbols when you decided to build, troubleshoot, or repair something? If so, you have the solution in your hands.

Don't give up on electronics when you encounter strange-looking circuit diagrams. You don't quit your favorite sport because you fear the rigors of training, do you? No! You get into condition with practice. Schematic dia-grams (or "schematics"), sensibly drawn and neatly arranged, can help you design, build, maintain, and repair electronic equipment. But you must do some work to gain skill at reading and interpreting schematics.

As you plan a trip by car, road maps show you how to navigate the coun-tryside. As you work with electronic equipment, schematics show you the way through simple circuits, complex devices, and massive systems. Once you know what the symbols represent, you'll find schematics no more difficult than road maps.

While you read this book, you'll learn the rationale of schematics, how to draw or interpret each symbol, and how the symbols interconnect to form functional circuits. You'll also get a chance to do a few simple experiments. Then you can continue your quest in any field of electronics from amateur radio to space communications, from surround sound to virtual reality.

You'll find my website at **www.sciencewriter.net**. I also create videos; simply search YouTube for my name. Have fun!

Stan Gibilisco

1

The Master Plan

You'll encounter three types of diagrams in electricity and electronics literature. Each style serves a unique purpose. When you buy an electric or electronic device or system, it should (in the ideal case) come with an operating and maintenance manual that includes all three types of diagrams.

- A *block diagram* gives you an overview of how the individual circuits in a system work together. You'll see each circuit represented as a "block" (rectangle or other shape, depending on the application). Interconnecting lines, sometimes with arrows on one or both ends, show how the circuits combine to form the whole system, and how currents and signals flow among those circuits. Figure 1-1 is a simple example.
- A *schematic diagram* (often simply called a *schematic*) shows every component in a circuit. Each component has its own special symbol. Lines between the components reveal how they connect together, and to a source of power, so they perform a specific function or operation. This book deals mostly with schematics. Figure 1-2 is a simple example.
- A *pictorial diagram* (sometimes called a *layout diagram*) shows the physical arrangement of the components on a circuit board or chassis so you can identify them for installation, testing, or replacement. Some such "diagrams" are actual photographs. Keep in mind, however, that pictures rarely reveal the electrical events that occur in a circuit or system. Figure 1-3 is a simple example.

1

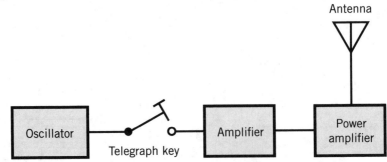

FIG. 1-1   *Block diagram of a radio transmitter that can send signals in Morse code.*

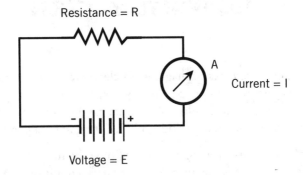

FIG. 1-2   *Schematic that includes a battery, a resistor, and an ammeter (labeled A).*

> **Tip**
> When you troubleshoot a malfunctioning system, you'll usually start with its block diagram to find the circuit in which the problem originates. Then you'll refer to the schematic (or part of it) to find the faulty component. A pictorial diagram, if available, will show you what the bad component looks like and where it resides within the system structure.

Block Diagrams

A block diagram can help you understand how a system works, and can help you troubleshoot it when it malfunctions. Each block has a label that describes or names the circuit it represents, but it doesn't explain the workings of the circuit, nor does it depict the individual components. When you gain a gen-

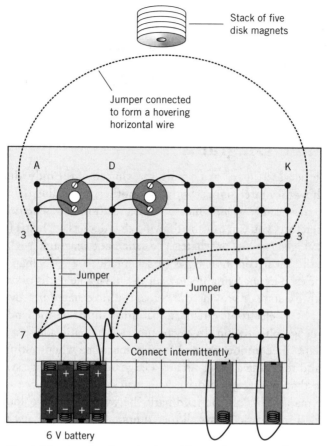

Stack of five
disk magnets

Jumper connected
to form a hovering
horizontal wire

A   D   K

3   3

Jumper

Jumper

7

Connect intermittently

6 V battery

FIG. 1-3 *Pictorial drawing that shows the physical layout of a test circuit for magnetism experiments.*

eral understanding of how a system operates by examining its block diagram, you can consult each of its circuit schematics for more details. Consider two examples.

- You want to design an electronic device to perform a specific task. You can simplify the process by drawing a block diagram that shows all the circuits you'll need to complete the project. Then you can expand each block into a schematic. In the end, you'll have a complete schematic that replaces all the blocks and shows the whole device in detail.
- Alternatively, you can approach the task the other way around. Imagine that you have a complicated schematic, and you want to use it to

troubleshoot a device. Because the schematic shows every single component, you might find it difficult to determine which part of the device has the problem. A block diagram can help you envision how each circuit works in conjunction with the others. Once you've found the troublesome circuit with the aid of the block diagram, you can examine its schematic and do tests to isolate the faulty component.

Schematic Diagrams

A schematic acts as a "map" of a circuit, showing all of the individual components and how they interconnect with one another. According to one popular dictionary, the term *schematic* means "of or relating to a scheme; diagrammatic." Therefore, you can call any drawing that depicts a scheme (electronic, physiological, geographic, or whatever) a schematic diagram.

One of the most common types of schematic is a road map for use in motor vehicles. The map might show all the navigable paths of travel inside a town, within a state or province, or across multiple states or provinces. Like a schematic of an electronic circuit, a road map shows all the landmarks relevant to a geographic region. In electronics, a schematic allows a technician to extrapolate the components and interconnections when testing, troubleshooting, and repairing a small circuit, a large device, or a huge system.

Suppose that you want to drive your truck from one place to another. Your road map shows all the landmarks between these two locations. By comparison, a schematic shows all the components between any two points in an electronic circuit. But both diagrams indicate more than mere points. You need to know more than which towns lie between two fixed locations to get an idea of the overall nature of the region. You could write down the names of the various towns or landmarks along a chosen route, but such a list couldn't take the place of a good road map. From an electronics standpoint, you could do the same thing by compiling a list of the components in certain circuit, such as:

- Two 120-ohm resistors
- One 1000-ohm resistor
- One PNP transistor
- Two 0.47-microfarad capacitors
- 90 centimeters of hookup wire

- One 6-volt "lantern" battery
- One switch with a built-in circuit breaker

This list tells you the "ingredients" of the circuit, but nothing in a functional sense. You know all the components necessary to build the circuit, but you don't know what it will do when you put it together! In fact, you might combine these components in several different ways to make circuits that do different things.

A schematic must not only show all the components in a circuit, but also how these components work with each other. A road map connects towns and other points of interest with lines that represent streets and highways. A line that indicates a secondary road differs from a line that represents a four-lane highway. With practice, you can learn to tell at a glance which sorts of lines indicate which types of roads. In electronics, a schematic uses a solid line to indicate a plain electrical conductor such as a wire or foil run; other types of lines (or sets of lines) represent cables, logical pathways, shielding enclosures, and wireless links. Whenever you draw an interconnecting line or set of lines, you portray some relationship between the connected components.

Schematic Symbology

A schematic uses *symbology* to reveal the anatomy of a system. On a road map, many of the symbols are lines to indicate roadways. But of course, a single black line that portrays State Route 522 doesn't resemble the appearance of this highway as you drive along it! You need only know the fact that the line symbolizes State Route 522. You can make up the other details in your mind. If you always had to see pictorial drawings of highways on paper road maps, those maps would take up thousands of times more space than the folded-up papers that you keep in your vehicle.

On a well-produced road map, you'll find a key to the symbols. The key shows each symbol and explains in simple language what each one means. If a small airplane drawn on the map indicates an airport and you memorize this fact, then each time you see the airplane symbol, you'll know that an airport exists at that particular site as shown on the map. Symbology depicts a physical object (such as an airport) in the form of another physical object (such as an airplane image).

Tip

In the few years since the previous edition of this book hit the presses, the Internet has evolved to the extent that you can use a computer, tablet device, or mobile phone to access road maps that show photographs of the views you'll see as you drive along some roads and highways. Many vehicles feature Global Positioning System (GPS) displays that show your location on a regular map (but beware: they're not always right!). You can look at photos taken from satellites, aircraft, and vehicles that have driven along specific routes. Check out "Google Maps," for example. These maps aren't printed on physical sheets of paper, so they aren't confined to the space limitations that paper maps impose. But they can distract you from the business of driving. I personally avoid them, or pull off the road and out of the way of traffic, before I look at them.

A road map contains many different symbols. Each symbol is "human engineered" to make sense in your mind. For instance, when you see a miniature airplane on a road map, you'll probably know that this location has something to do with airplanes, so you won't need a detailed explanation. If, on the other hand, the mapmaker used a beer bottle to represent an airport, anyone who failed to read the key would probably think of a saloon, not an airport! Because a map needs many different symbols, a good mapmaker tries to make sure that the symbols make sense.

Logical thinking will only take you to a certain point in devising schemes to represent complicated things, especially when you get into the realm of electronic circuits and systems. For example, a circle (or sometimes an ellipse) normally forms the basis for a transistor symbol, a silicon-controlled rectifier (SCR) symbol, and an electric outlet symbol. Additional symbols inside the circle reveal which type of component it represents.

In the olden days of electronics, engineers used circles containing various electrode symbols to represent vacuum tubes. (Sometimes they still do!) Transistors have evolved to replace vacuum tubes in most situations, so the schematic symbol for a transistor also starts with a circle. Electrode markings go into the circle as before, but transistor elements differ from tube elements, so transistor circles contain different markings than tube symbols did. Transistors perform many of the same functions as vacuum tubes did (and sometimes still do!) so their symbols look somewhat alike, but they're far from identical.

Inconsistencies arise in schematic symbology, so electronics-related diagrams can get a lot more sophisticated and subtle than any road map you'll ever see. For example, you can portray an SCR as a circle with a diode symbol inside and an extra line coming out of it. But an SCR performs a function that differs from what a tube or transistor does. An electric outlet can serve as another example. It doesn't work anything like a tube or transistor or SCR, but the basis for the symbol is a circle, just like the circle for a tube or transistor or SCR. You'll learn more about schematic symbols in Chapter 3.

Variations on a Theme

Sometimes you'll see a circle-free schematic symbol for a component that normally includes a circle. This non-standard style appears fairly often with the symbols for diodes, transistors, and SCRs. (It almost never happens with the symbols for vacuum tubes or electric outlets.) Figure 1-4 shows a PNP bipolar transistor symbol with the usual circle (A) and without it (B). Don't let "mutants" like this confuse you! The symbols differ but the components are identical.

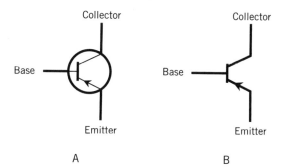

A   B

FIG. 1-4   *At A, the schematic symbol for a PNP bipolar transistor including a circle. At B, the same symbol without the circle.*

Component Interconnections

Consider a single, commonplace electronic component: a PNP bipolar transistor. This device has three electrode elements, and although many different varieties of PNP bipolar transistors exist, you draw all their symbols in pretty much the same way. You might find a PNP bipolar transistor in any one of thousands of different circuits! A good schematic will describe:

- How that particular component fits into the circuit.
- Which other components work in conjunction with it.
- Which other circuit elements depend on it for proper operation.

A PNP bipolar transistor can act as a switch, an amplifier, an oscillator, or an impedance-matching device. If a PNP bipolar transistor functions in some circuit as a *radio-frequency* (RF) amplifier, you can't conclude that a PNP bipolar transistor can operate as an RF amplifier only, and nothing else. For example, you could pull the thing out of the RF amplifier circuit and put it into another circuit to serve as the "heart" of an *audio-frequency* (AF) oscillator.

Tip

If you know the identity of a component (say, a PNP transistor) but nothing else about it, you can't tell what role it plays in a circuit until you see a schematic that shows all the components in the circuit and how they interconnect. Rarely can you get all this information in easy-to-read form by examining the physical hardware. You need a "road map" (schematic) to show you all the connections that the engineers and technicians made when they designed and built the circuit.

Imagine that you decide to drive your car from Baltimore, Maryland to Los Angeles, California. Even if you've made the trip several times in the past, you probably don't recall all the routes that you'll need to take and all the towns and cities that you'll pass along the way. (They might have changed, anyway, if you haven't made the trip for some time!) An up-to-date road map will give you an overall picture of the whole trip. Because all the trip data exists in a form that you can scan at a glance, the road map plays a critical role in allowing you to envision the entire trip rather than each and every piece, one at a time. A schematic does the same thing for a "trip" through an electronic circuit. And, like road maps, schematics can evolve over time as engineers make improvements to the original design.

Continuing with the road map and the coast-to-coast trip as an example, imagine that you've memorized the entire route from Baltimore to Los Angeles. Assume also that one of the prime highways along the way has gone under construction, forcing you to take an alternate route. Without a road map, you'll have no idea what other roads exist in that area, which alternate route to take, and which detour will keep you on course as much as possible and eventually return you to the original travel route with a minimum of delay and inconvenience.

An electronic circuit has many "highways" and "byways." Occasionally, some of these routes break down, making it necessary to seek out the problem and correct it. Even if you can visualize the circuit in your head, you'll find it impossible to keep in your "mind's eye" all the different electrical paths that exist, one or more of which could prove defective. When I write of visualizing the circuit, I don't mean the schematic equivalent of the circuit, but the circuit's actual components and interconnections, known as the *hard wiring*.

A schematic gives you an overall map of a circuit and shows you how the various routes and components interact. When you can see how the complete circuit depends on each individual route and component, you can diagnose and repair any problem that might arise. Without such a view, you'll have to "shoot in the dark" if you want to get a malfunctioning circuit working again. You might even introduce a new problem instead of resolving the original one!

Fear Not!

Look at the schematic of Fig. 1-5. If you've had little or no experience with these types of diagrams, you might wonder how you'll ever manage to interpret it and follow the flow of currents and signals through the circuit that it represents. Fear not! When you finish this book, assuming that you knew some basic electricity and electronics principles to begin with, you'll wonder how you could ever have let a diagram like this intimidate you. (By the way, you'll see this diagram again in Chapter 5.)

A Visual Language

Every word in English or any other verbal language is a complex symbol made from simpler elements called *characters*. Let's take the word "stop," for example. Without a reference key, this sound means nothing. A newborn infant hears noise coming out of your mouth, that's all! But through learning the symbology, this word acquires meaning because the child, who has begun to speak and understand, can compare "stop" to other words, and also to actions. You can even say that the word "stop" is a sort of symbology within symbology. Your intent, when using the word "stop," can also be expressed by the phrase "Do not proceed further." This phrase also constitutes symbology, expressing a mental image of a desired action.

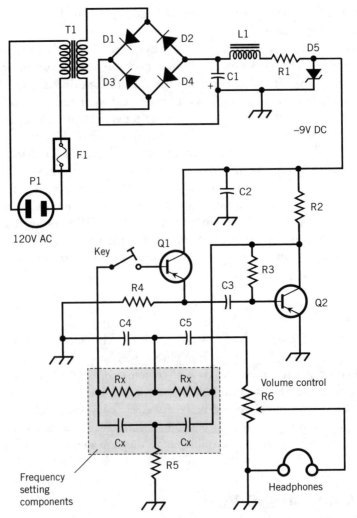

FIG. 1-5   *A rather complicated schematic. When you've finished this book, you'll think it's simple! (Don't worry about the component designators right now.)*

If people could communicate by mental telepathy, then no one would need language or the symbols that it comprises. Thinking happens faster than anyone can speak or read or write. Brain processes are the same from human to human, regardless of what language any particular person employs when speaking, reading, or writing. A newborn baby speaks and understands no language. But whether that baby was born in America, South Africa, China, India, or wherever, the same thought processes take place.

A baby knows when it's hungry, in pain, frightened, or happy. It needs no language to comprehend these states. But the baby does have to communicate right from the start. For this reason, all newborns communicate in the same language (crying and laughing, mostly). As babies comprehend more of their environment through improved sensory equipment (eyes, ears, nose, fingers), they collect more data. Then the various languages come into play, with different societies using different verbal symbols to express mental processes. The human brain nevertheless carries on the same nonlinguistic thought processes as before, because thinking in terms of symbols alone would take too much time and "brain storage."

The brain helps a human to transpose complex thoughts into language and vice-versa, just as a computer translates programming languages into electronic impulses and vice-versa. Imagine that a child is about to step in front of a speeding automobile. If your brain had to handle millions of data elements symbolically, you would spend a lot of time waiting for your brain to deliver the correct processed information, and that child would probably get killed before you could take any action. Rather, your brain scans all the data received by your sensory organs in quick time and then sums it up into a single symbol for communication. A good audible (and hopefully loud) symbol in the above-mentioned case is "Stop!" You, seeing a child about to walk into heavy traffic, might shout that word and produce in the child's brain the appropriate sequences of processes.

Not all languages involve spoken words. You've doubtless heard of sign language, whereby a person's arms and hands move to communicate ideas. If you've done any amateur ("ham") radio communication, especially if you got your "ham" license back in the time when I got mine (the 1960s), you know the Morse code as a set of communication symbols. In most instances, a language comprising only visual symbols or audio symbols is not as efficient for us humans as one composed of visual and audio symbols combined. Using the symbol "stop" again, you can utter this word in many different ways. The word in itself means something, but the way we say it (our "tone of voice") augments the meaning. You can't do all that with the printed or audible characters S, T, O, and P in plain text or in Morse code.

We humans have arrived at universal methods of modifying visual symbols. For example, we often use color in conjunction with the visual symbol for a spoken word. Think of a "stop" sign. It's red, right? People tend to associate red with the word "stop" or "danger." Or think of a "yield" sign. It's yellow, representing something that demands attention, but in a less forceful way than red does; you "proceed with caution." When a traffic light turns green, you can "go!" (But considering how some fools drive nowadays, you'd better use caution all the time if you want to live very long.)

> ### Tip
> Schematic diagrams rarely include color. Look in the back of a technical manual for an amateur radio receiver or transmitter. Does the schematic have color? I'll bet that it doesn't. (A few high-end magazines do, though.) The schematic in the back matter of a technical manual might not even have grayscale shading. Schematics resemble printed text or Morse code in this respect; engineers must convey a lot of information with a limited set of symbols, and they're constrained even as to the way in which they can portray those symbols.

Schematics don't lend themselves to any form of oral (audible) symbology either. When you see the symbol for, say, a field-effect transistor (FET) in a schematic diagram, you don't hear the paper or computer say, "Field-effect transistor, for heaven's sake, not bipolar transistor!" You have to make sure that you read the symbol correctly. If you want to build the circuit and you mistakenly put a bipolar transistor where an FET should go, then you can't expect the final device or system to work. Something might burn out, so that when you recognize your error and bipolar transistor with an FET, you'll have to troubleshoot the whole circuit before you can use it. You might even have to start all over again and replace every single component!

Your senses along with your central processor (your brain) render you less than proficient at mentally conceiving all the workings of electronic circuits by dealing with them directly. You must accept data one small step at a time, compiling it in hardcopy form (through symbology) and providing a hardcopy readout. You can liken this method to "connect-the-dots" drawings in children's school workbooks. Individually, the dots mean nothing, but once they're arranged in logical form and connected by lines, you get an overall picture. The dots' relationships to each other and to the sequence in which they're connected tell you all you need to know.

The remaining chapters in this book start with the symbols for individual electronic components, then move on to simple circuits, and finally show you a few rather complicated circuits. Schematic symbols and diagrams are designed for humans, so human logic plays a prime role in determining which symbols mean which things. In that respect, the creation and reading of schematic diagrams resembles mathematics, and in particular, old-fashioned plane geometry!

Homogenize This!

Schematics comprise encoded representations of circuits, while pictorials show you the physical objects, often proportioned according to their relative size, and sometimes rendered to appear three-dimensional by means of shading and perspective. Schematics depict circuit components as symbols only, without regard to their real-world size or shape, and in two dimensions (on a flat piece of paper or computer screen), lacking depth or perspective. Nevertheless, in a few short decades, your computer will create a three-dimensional schematic or pictorial hologram within which you can walk around and see components to your left, to your right, in front of you, behind you, over your head, under your feet, and maybe (gulp) inside your body.

2

Block Diagrams

A block diagram portrays the general structure of a device or system. Such a diagram can provide a simplified rendition of a complicated system by separating its main parts and showing you how they interconnect and interact. In computer engineering, block diagrams can help you envision how programs or other processes work.

A Simple Example

Figure 2-1 shows a block diagram of a device that converts *alternating current* (AC), of the sort you find at the electric outlets in your house, to *direct current* (DC), of the sort you get from an electrochemical battery. Hobbyists and professionals call this type of device a *power supply*.

The terminal at the far left accepts the AC input. As you go from left to right, the electricity passes through the transformer, the rectifier, and the filter before arriving at the output as *pure DC*. In this case, the lines between blocks have no arrows; the diagram's creators assume that you can sense the process direction without them.

> **Tip**
> In some block diagrams, the interconnecting lines include arrows to clarify which block affects which, or to indicate the general direction of signal flow when you might not sense it by instinct.

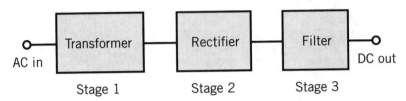

FIG. 2-1 *Block diagram of an AC-to-DC converter, also called a power supply. You'll naturally sense that the electricity flows from left to right, so the lines have no arrows.*

Functional Drawings

Block diagrams can indicate the interconnections among small circuits in a larger device, or among diverse devices in a massive system. When you see a block diagram rendered in the style of Fig. 2-1, you can call it a *functional diagram* because it tells you something (but not a lot) about what the device does. If you want to know more detail, you'll need to see the schematic.

An engineer who wants to design a complex electronic system can start with a block diagram. It shows all the circuit sections (stages) in a functioning device, but none of the internal details of those stages. Then the engineer develops schematics of circuits that fill each block and serve the appropriate function. The first block gets replaced by the schematic of the circuit it represents. The engineer proceeds through the blocks in functional order, creating schematics that you can use to build each stage in the system. When every block has been replaced with a schematic, a detailed (but so far only theoretical) system design exists.

Another way of using block diagrams involves starting with the complete schematic of a system. Imagine that the schematic is quite complicated, and for some unknown reason the system doesn't work as the engineer thinks it should. Although a schematic can describe the functioning of an electronic system, it's not as clear as a functional block diagram for that purpose. The schematic literally has *too much* information! Lacking a block diagram, a repair technician would have to start with the schematic, laboriously identify each stage in the system, and then draw the entire system diagram in block form. When finished, the block diagram would reveal how each stage interacts with the others. Using this method, the technician could identify one or more stages as likely trouble zones, and refer back to the original schematic to conduct tests in those suspect circuits.

Tip

Even when presented without accompanying schematics, a block diagram can tell you the functional operation of an electronic system. The block diagram can come in handy when you don't need to know all the details about what every single component does, but only the general way that the system works.

You can describe the operation of a specific type of radio system, for example, an *amplitude-modulated* (AM) voice transmitter, by means of a block diagram. Of course, no two AM transmitters built by different manufacturers are identical, but all of them contain similar functional stages. One type of oscillator might work differently from another type, but all oscillators do the same thing: generate an RF signal! When you want to know or portray small differences among circuits that do essentially the same things, then you need schematics of them all.

Believe It or Not!

Although "plain old AM" is technically obsolete, some radio broadcast stations and *Citizens Band* (CB) radio transmitters still use it, and will probably keep on using it for decades to come!

The block diagram of Fig. 2-2 shows a strobe light system as you might see it in the assembly manual for a kit comprising self-contained circuits, a set of cables, and an instruction manual. You connect the cables (solid lines with arrows) among the self-contained circuits (blocks) provided with the kit, meticulously following the instructions in the manual.

The input signal enters at the left; it's utility AC such as you get from a standard wall outlet. In the United States, this AC has a nominal voltage of 117 volts (117 V) and a frequency of 60 hertz (60 Hz), where "hertz" means "cycles per second." (In some countries the voltage is about 234 V, and in some countries you'll find a frequency of 50 Hz rather than 60 Hz.) The input AC goes to a fuse, and also to a combination of components that provide timing.

The top path, where you see the fuse, leads to a rectifier whose output passes to one terminal of a three-terminal strobe lamp. The rectifier output also connects to an adjustable timer that provides a variable flash rate for

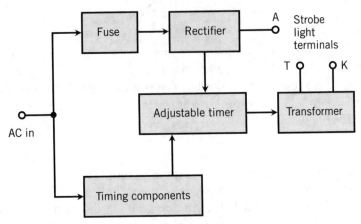

FIG. 2-2 *Block diagram of a circuit designed to provide power to a strobe light. Arrows show how the electricity flows.*

the lamp. The timer output goes to a transformer, which in turn connects to two more lamp terminals. You don't have to know what the designators "A," "T," and "K" mean; you need only know how to hook up the cables as you assemble the kit! Some people call this sort of "monkey see, monkey do" instructional drawing a *wiring diagram*.

Current and Signal Paths

Figure 2-3 is a block diagram of a power supply that produces several outputs having various electrical characteristics. As you proceed through the diagram from the left-hand end (the input) to the right and downward (the outputs) according to the arrows, you'll see that the system operates from 117 volts AC (117 VAC), commonly found at utility outlets in the United States. The entire power supply could reside in a single cabinet with a single cord and plug for a wall outlet and screw-on terminals for connection to multiple devices.

The input AC gets split into two identical paths, both at 117 VAC. One splitter output goes to the "lower" transformer that provides 16 VAC and 3 VAC output. The other splitter output runs to the "upper" transformer, which goes to:

- A "top" rectifier/filter that provides +12 volts DC (+12 VDC) without voltage regulation
- A power "off" detector, such as an AC voltmeter or alarm

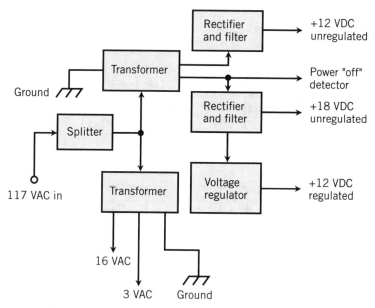

FIG. 2-3 *Block diagram of a power supply that produces several different outputs.*

- A "bottom" rectifier/filter that provides +18 VDC without voltage regulation

In addition, the "bottom" rectifier/filter output connects to a voltage regulator that maintains it at a steady +12 VDC regardless of minor power surges and dips in your household voltage as it comes from the utility company.

Did You Know?

Most voltage regulators work by brute-force limiting. They take DC electricity at a certain voltage (say, +18 VDC) and hold it down so that it can't exceed a certain lower voltage (say +12 VDC). This methodology can go only so far, however. If the utility electricity dips to 80 or 90 VAC, the regulator of Fig. 2-3 will probably manage to keep the output at +12 VDC. But a utility dip to only 40 VAC would likely cause the "regulated" voltage to fall from +12 VDC to something like +6 VDC.

Figure 2-4 is a block diagram of a simple AM radio transmitter. You speak into the microphone, which leads to an AF preamplifier to give your voice

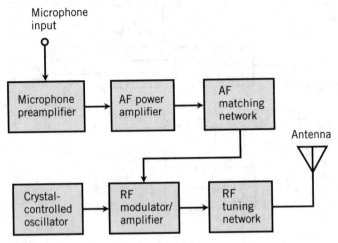

FIG. 2-4 *Block diagram of an AM radio transmitter.*

signal some power (but not much). A second AF amplifier gives your voice a lot of "punch"! The AF matching network ensures that the voice signal will deliver the most possible power to the modulator/amplifier, which receives its RF energy from an oscillator whose frequency is determined by a quartz crystal. The instantaneous RF modulator/amplifier output power fluctuates in accordance with the instantaneous AF input level to produce an AM signal, which passes through an RF tuning network to the antenna.

Aha!

Figure 2-4, with its arrows, tells you not only how the system components connect to each other, but also the sequence of events and the directions in which the AF and RF signals flow from the microphone and oscillator to the antenna.

Flowcharts

Block diagrams can describe how electronic systems work, but in the world of computers, another form of diagram, called a *flowchart*, can portray the functioning of a program or *software*. A flowchart resembles a block diagram, except that the symbology applies to the sections of a computer program, an intangible thing (as opposed to an electronic system, a tangible thing). A flowchart provides a graphic representation of the logical steps that a computer takes as it executes a program. Software engineers prepare flowcharts in conjunction with specifications, and modify the flowcharts as user requirements change.

For complex problems, a formal written specification can ensure that everyone involved understands and agrees on the nature of the problem, and on the desired results of the program. For example, suppose that a schoolteacher (that's you!) uses a computer program to help determine a student's final course grade by calculating an average from scores the student got for quizzes during the course period. You input each and every quiz score to the program. The program outputs the average of all those scores. Figure 2-5 shows a flowchart of the program process, as follows.

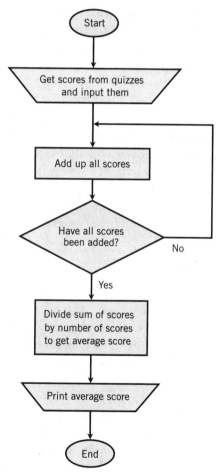

FIG. 2-5 *A flowchart that describes a computer program.*

- Receive the quiz scores from you.
- Add up all the quiz scores to get their sum.
- Verify that you have accounted for all the quiz scores.
- Divide the sum by the number of quizzes to get the average quiz score.
- Print the average quiz score.

The computer leaves the hardest task in the process to you, the teacher: Decide what grade the student deserves! If the quizzes were difficult, you might accept fairly low average scores for standard "letter grades" (such as A, B, C, D, or F); if the quizzes were easy, you might demand higher average scores for given grades.

The flowchart graphically presents the structure of the program, revealing the relationship among the steps and paths. When a program has many different paths that result from numerous decisions, a flowchart can help you sort things out. You can use the flowchart as a tool to understand the problem and to aid in program design.

Tip

Flowchart symbols contain narrative descriptions rather than programming language statements, because you want to describe what happens, not how it happens. Later, if you want to create flowcharts for documentation, they can contain statements in a programming language. These flowcharts might prove helpful to another person who at some future time wants to understand or modify the program.

You might need quite a lot of time to conceive and draw up a good formal flowchart. Modifying a flowchart to incorporate changes, once a program has been written and its flowchart composed, can prove difficult. Because of these limitations, some programmers shy away from flowcharts, but for others they provide valuable assistance in understanding a program. In order to promote uniformity in flowcharts, standard symbols have been adopted. Figure 2-6 shows the most common ones. In a sophisticated flowchart, you might find them all.

Ovals show start or stop points. Arithmetic operations go in rectangular boxes. Input and output instructions go in trapezoids. If you want to show a program that someone wrote earlier within the context of a larger flowchart, you don't necessarily have to draw the flowchart for the entire "subprogram."

Instead, you might represent the entire program as a flattened hexagon. If a box indicates a decision, you use a diamond shape. A five-sided box portrays a part of the program that changes itself. A small circle identifies a processing junction point. Such a point in the program can go to several places. A small five-sided box, which has the shape of the home plate on a baseball field, shows where one page of a flowchart connects to the next, if the entire flowchart has more than one page.

You should label all intermediate junction and off-page connection points with numbers and letters to tell your readers that all like symbols with the same character inside connect together. Arrows indicate the direction of the flow.

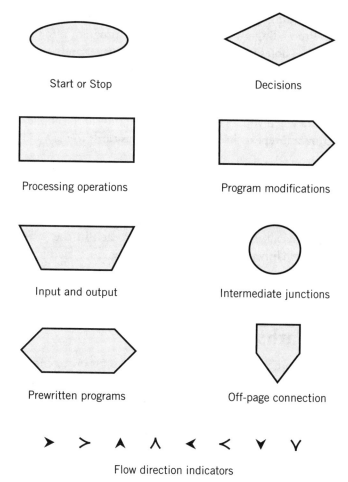

FIG. 2-6 *Common symbols for flowcharts intended to represent computer programs.*

Down and to the Right

The normal direction of processes in a flowchart runs from top to bottom and from left to right, the same way as people read books in most of the world. Arrowheads on flow lines indicate direction. You can omit the arrows if (but only if) the direction of flow is obvious without them.

Figure 2-7 shows a flowchart for a program that duplicates punched cards, and at the same time prints the data on each card. Let's trace the flow. The program begins at the "Start" oval at the top and proceeds in the direction of the arrows. According to the text in the trapezoid below "Start," the program reads a card. Proceeding on down the chart, the program punches the card's contents (data) as holes in a blank piece of heavy paper and sends the data to a printer. The program then goes back along the dashed line to the top and reads the next card. The circles marked "A" represent inflow and outflow points. In this case they're superfluous, but in a complicated flowchart they can prove useful when you'd get a mess by including all the applicable dashed lines. The program repeats itself as long as it has cards to read and punch.

Old but Good

The foregoing program makes a good history lesson! Were you born long enough ago to remember punch cards for inputting programs to computers? I recall using them, all the way back in the 1970s, when I attended the University of Minnesota. That little factoid dates me, doesn't it?

Process Paths

Let's look some more at Fig. 2-7. Suppose that you want to change the card-punching program so that the computer ignores blank (hole-free) cards and duplicates only those cards that have at least one hole. Because the computer must make a decision about each card, you'll need to include a decision block in the flowchart. Figure 2-8 shows the result.

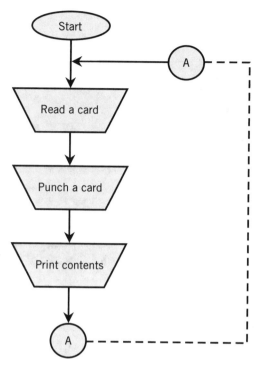

FIG. 2-7 *Flowchart that outlines the steps in a program intended to duplicate punched cards. The circles labeled "A" represent inflow and outflow points in the feedback loop shown by the dashed line.*

Follow the Flow

Except for the decision block, Fig. 2-8 shows the same process as Fig. 2-7 does. The program begins in the "Start" oval at the top and then goes to the block marked "Read a card." From there, the program moves on to the decision block labeled "Card blank?" If the answer is "Yes" (the card has no holes), the program proceeds to the connection circle marked "A" and back to the top to read the next card. If the answer is "No" (the card has at least one hole), the program instructs the hardware (the physical components of the computer and its peripherals) to punch a duplicate card and print its contents. Then the program goes to another circle marked "A" and back to the starting point.

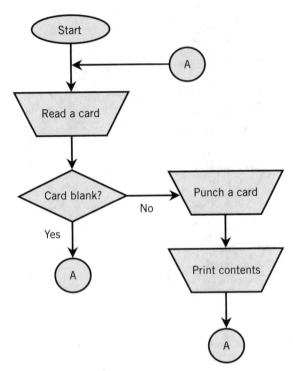

FIG. 2-8 *A flowchart with a decision block (diamond). The circles labeled "A" all represent a single junction point through which data moves as shown by the arrows.*

Tip
Figure 2-8 is a simple flowchart, showing a process that uses only input and output devices and performs no calculations. Most programs and flowcharts involve more complicated processes.

Microcomputers use many different types of diagrams that deal mostly with software (operating systems and programs) rather than hardware (physical components). In the computer world, functional block diagrams abound and are usually more numerous than schematic diagrams. From an understanding standpoint, block diagrams can serve to portray machine functions in general, but hardware maintenance and repair procedures require well-defined schematic drawings. Computers take advantage of the latest state-of-the-art developments in electronic components and are relatively simple when you consider all the things they can do. However, from a pure electronics standpoint and as far as schematic diagrams are concerned, computers are immensely complicated. You'd need a lot of pages full of schematics to represent even the most rudimentary computer.

Summary

Block diagrams can help you show and understand how electronic circuits work. They're comparatively easy to draw, usually requiring only a marking instrument, some paper, and a straight edge (or a vector-graphics computer program and a little bit of training on it). Schematic diagrams, in contrast, need more tools and can, in some cases, take many hours to render in a form that people can read and interpret.

3

Components and Devices

On a road map, symbols indicate geographical features such as cities, highways, airports, railroad tracks, and other landmarks. The same rule applies to schematics in electricity and electronics. Specialized symbols portray conductors, switches, resistors, capacitors, inductors, transistors, and other circuit elements. Whenever engineers invent a component or device, they create a new schematic symbol for it.

Tip
In this chapter, you'll see common schematic symbols for some (but by no means all) of the components that you'll find in electrical and electronic systems. In the back of this book, Appendix A provides a more comprehensive listing.

Resistors

Resistors rank among the simplest electronic components. As the term implies, they resist or impair the flow of electric current. Engineers express *resistance* (the extent of current impairment) in units called *ohms*. Most real-world resistors have values ranging from approximately 1 ohm up to millions of ohms. Once in a while, you'll encounter resistors with values less than 1 ohm, or values in the thousand-millions (billions) or million-millions (trillions) of ohms.

Regardless of their ohmic value, nearly all *fixed resistors* have schematic symbols that look like Fig. 3-1A or B. The two horizontal lines at the left and right (A) or the top and bottom (B) depict wires called *leads* that protrude from the ends of the physical component. Some resistors have rigid metal terminals such as pins or lugs that don't necessarily come out of the ends.

Figure 3-2 shows a "transparent" pictorial of a *carbon-composition* fixed resistor with wire leads on both ends. Figure 3-3 shows pictorials of two other types of resistors: *wirewound* (A) and *film* (B). You can denote any resistor of the sort shown in Fig. 3-2 or Fig. 3-3 with either of the symbols in Fig. 3-1.

FIG. 3-1   *Symbol for a fixed-value resistor. In a schematic, it can appear horizontal (A) or vertical (B).*

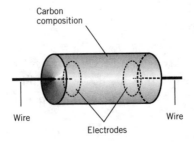

FIG. 3-2   *"Transparent" pictorial of a carbon-composition resistor.*

Tip

You can rotate almost any schematic symbol by 90 degrees, as in Fig. 3-1, to make it fit in a diagram. You can even turn it upside down or backwards if necessary! If your viewers know what a symbol means, its orientation doesn't matter.

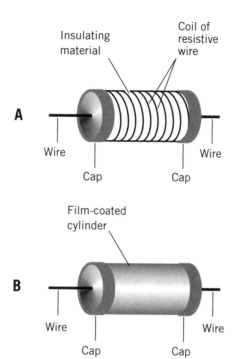

FIG. 3-3 *Pictorials showing the anatomy of a wirewound resistor (A) and a film type resistor (B).*

A *variable resistor* has an ohmic value that you can adjust by moving a slide or tap along the resistive element. You set the resistance to a specific value, where it remains until you deliberately change it. The circuit "sees" the component as a fixed resistor at any given time.

When a circuit contains a variable resistor, the schematic reveals that fact. Figure 3-4 shows a common symbol for a variable resistor with two terminals. Some variable resistors have three terminals. Figure 3-5 shows two examples of schematic symbols for a three-terminal variable resistor known as a *potentiometer* or *rheostat*, depending on the method of construction.

FIG. 3-4 *Symbol for a two-terminal variable resistor.*

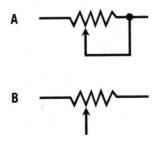

FIG. 3-5 *Symbols for three-terminal variable resistors, also known as potentiometers or rheostats (depending on the method of manufacture). At A, the center terminal connects to one end terminal to obtain, in effect, a two-terminal component. The resistor at B has three independent terminals.*

Did You Know?

A rheostat contains a wirewound resistance element, while a potentiometer is normally of the carbon-composition or carbon-film type. You can vary a rheostat's value in small increments or steps, but you can adjust a potentiometer's value over a continuous range. Rheostats contain inductance along with resistance, while potentiometers have pure resistance with essentially no inductance.

Tip

In a schematic symbol, an arrow sometimes indicates that a component has variable or adjustable value. But not always! The symbols for transistors, diodes, and some other solid-state devices contain arrows that have nothing to do with variable or adjustable properties.

Figure 3-6 shows a variable resistor of the wirewound type, manufactured to expose an uninsulated coil of resistance wire. You can adjust a sliding metallic collar, which goes around the body of the resistor, to intercept different points along the coil. A flexible conductor connects the collar to one of the two end leads. The collar shorts out more or less of the coil turns, depending on where it rests along the length of the coil. As you move the collar to the right along the wire coil, the ohmic value between the two end leads decreases.

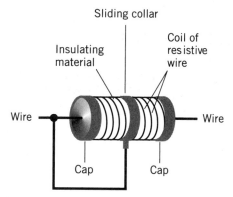

FIG. 3-6 *Pictorial of a wirewound variable resistor with the movable middle sleeve connected to one of the fixed end leads.*

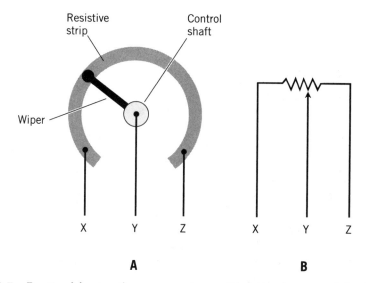

FIG. 3-7 *Functional drawing of a rotary potentiometer (A) and its schematic symbol with corresponding connections (B).*

Figure 3-7A is a functional drawing of a rotary potentiometer. Figure 3-7B shows its schematic symbol. The symbol has three distinct contact points. When you rotate the control shaft, the resistance varies between the center contact and the end contacts. Figure 3-8 is a pictorial of a typical real-world potentiometer.

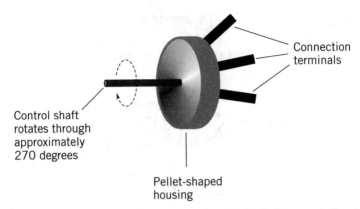

Connection terminals

Control shaft rotates through approximately 270 degrees

Pellet-shaped housing

FIG. 3-8 *Pictorial of a potentiometer suitable for mounting on the front panel of an electronic system such as a radio receiver.*

Tip

You can convert the variable wirewound resistor shown in Fig. 3-6 to a three-terminal component by severing the connection between the collar and the left-hand end. Then you can independently vary the resistance between the collar and either end. Likewise, you can convert a three-terminal wirewound resistor to a two-terminal component by shorting the variable contact point to either end.

The schematic symbol for a resistor, all by itself, says nothing about its ohmic value, or anything else about the component such as its power rating or physical construction. You'll often see specifications for the component written alongside the resistor symbol, but these details might instead appear in a separate "components list" and referenced to an alphabetic/numeric designation printed next to the schematic symbol (such as R1, R2, R3, and so on).

Tip

You can usually determine the ohmic value of a fixed resistor by looking at the colored bands or zones on it. Appendix B lists color codes that tell you the ohmic values of fixed resistors.

Capacitors

Capacitors are electronic components that can block direct current (DC) while passing alternating current (AC). They can also store energy in the form of an electric field. The basic unit of capacitance is the *farad* (symbolized F). One farad represents a huge electrical quantity, so most real-world capacitors are rated in tiny fractions of a farad: *microfarads* or *picofarads*. One microfarad (symbolized μF) equals a millionth of a farad (0.000001 F). One picofarad (symbolized pF) equals a millionth of a microfarad (0.000001 μF) or a trillionth of a farad (0.000000000001 F).

Figure 3-9 shows the most common symbol for a fixed capacitor. The curved side should go to electrical ground, or to the circuit point more nearly connected to electrical ground. On occasion, you'll see alternative symbols such as those in Fig. 3-10A or B.

Many types of capacitors exist. Some are *nonpolarized* devices, meaning that you can connect them in either direction and they'll work equally well. Others are *polarized*, having a positive and a negative lead or terminal. You must connect such a capacitor so that any DC voltage across it has the correct polarity.

Tip

Most capacitors have two leads or terminals, but every now and then you'll see one with three or more. You should check the technical documentation for such a capacitor so that you know where each lead or terminal should go in a circuit that contains it.

FIG. 3-9 *Standard symbol for a fixed capacitor. The curved line represents the plate (or set of plates) that's electrically closer to ground.*

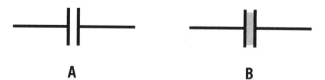

A   **B**

FIG. 3-10 *Alternate symbols for fixed capacitors. At A, air dielectric; at B, solid dielectric.*

Unless the symbol includes a polarity sign, it indicates a nonpolarized capacitor, which can have metal plates surrounding ceramic, mica, glass, paper, or other solid nonconductive material (and sometimes air or a vacuum). The nonconductive material, known as a *dielectric*, separates the metal parts of the component. A typical fixed-value capacitor comprises two tiny sheets (or sets of sheets) of conductive material that lie physically close to each other but are kept electrically apart by the dielectric layer.

Figure 3-11 shows the symbol for a polarized capacitor. It looks like the symbol for a nonpolarized capacitor, but a plus (+) sign appears on one side. The plus sign tells you that the positive terminal of the component should go to the more positive part of the external circuit. Occasionally, a minus (−) sign will appear on the opposite side instead of, or in addition to, the plus sign. The minus sign indicates that the negative terminal of the capacitor should go to the more negative part of the external circuit.

Beware!

Never connect a polarized capacitor the wrong way around. That mistake can damage the component. In the extreme, it can literally (maybe even violently) explode!

In this chapter, all the capacitors that you've seen thus far have a fixed design. In other words, the components have no provision for changing the capacitance value, which the manufacturer determines at the factory. But you can adjust the values of some capacitors at will. They're called *variable capacitors*. Some specialized types are known as *trimmer capacitors* or *padder capacitors*.

Figure 3-12 shows the most common symbol for a variable capacitor. An arrowed line runs diagonally through it. Figures 3-13A and B show two alternative ways of denoting the same component. All three symbols indicate that you can adjust the capacitance at will, regardless of the physical construction details.

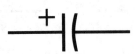

FIG. 3-11 *Symbol for a polarized capacitor. The side with the plus sign (+) should carry a positive DC voltage relative to the other side.*

FIG. 3-12 *Standard symbol for a variable capacitor. The curved line represents the rotor, and the straight line to its left represents the stator.*

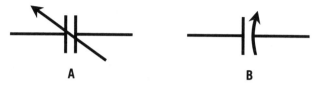

A   **B**

FIG. 3-13 *Alternate symbols for variable capacitors. At A, the stator is not distinguished from the rotor; at B, the rotor appears as a curved line with an arrow.*

An *air variable capacitor* (one with an air dielectric) allows you to tune many types of RF equipment including antenna matching networks, transmitter output circuits, and old-fashioned radios. A typical "air variable" has interlaced plates connected together alternately to form two distinct contact points. The rotatable set of plates is called the *rotor*; the stationary set of plates is called the *stator*. All variable capacitors are nonpolarized components, meaning that you can apply an external DC voltage either way and the performance remains the same.

> **Tip**
>
> In most air variables, the rotor should go to the electrical ground, regardless of the polarity of any external DC voltage that you apply. The rotor connects physically to the shaft that you turn. When you ground the shaft along with the rotor, you minimize *external capacitance* effects. If you bring your hand near a grounded shaft, your *body capacitance* doesn't upset the circuit's performance. Also, you avoid the risk of electric shock if you forget to put an insulated knob on the end of the shaft and touch it when you want to adjust it!

Sometimes you'll see multiple variable capacitors connected together or *ganged*. In a set of ganged variable capacitors, two or more components can control two or more circuits at the same time. The rotors, although physically separate, all share a single shaft. Figure 3-14 shows the schematic symbol for

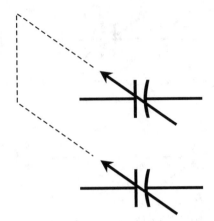

FIG. 3-14   *Symbol for two variable capacitors ganged together.*

two variable capacitors ganged together. The individual components might have identical minimum and maximum capacitance values, but not necessarily. In any case, they track together. When one capacitor increases in value, the other (or others) also increase in value.

> **Tip**
> The schematic symbol for a capacitor serves only to identify it and tell you whether it's fixed or variable, and if it's fixed, whether or not it's polarized. The component value might appear beside the symbol in the schematic, or the symbol might include a letter and number designation (for example, C1, C2, C3, and so on) for reference to a "components list" that accompanies the diagram.

Inductors and Transformers

A basic *inductor* comprises a coiled-up length of wire that introduces *inductance* into a circuit. Inductance opposes changes in electric current. With constant DC, an inductor stores electrical energy but offers no opposition to the current itself. Inductors can range in physical size from microscopic to gigantic, depending on the inductance value of the component, and on the amount of current that it can handle.

The standard unit of inductance is called the *henry* (symbolized H). That's a large electrical quantity. You'll find most inductors rated in *millihenrys* (symbolized mH), where 1 mH = 0.001 H, or in *microhenrys* (μH), where 1 μH = 0.001 mH = 0.000001 H. Occasionally, you'll see an inductor rated in *nanohenrys* (nH), where 1 nH = 0.001 μH = 0.000000001 H.

That's Weird!

The above spellings aren't typos. Most engineers write "henrys" rather than "henries," "millihenrys" rather than "millihenries," "microhenrys" rather than "microhenries," and "nanohenrys" rather than "nanohenries."

Figure 3-15 shows the schematic symbol for an *air-core inductor*. The two leads or terminals are designated by straight lines that merge into the coiled part. An air-core coil has nothing inside the windings that can affect the inductance. Some air-core coils are wound from stiff wire and support themselves mechanically. In other cases, a rigid form made out of plastic or ceramic material supports the coil turns, keeping them in place and enhancing the physical ruggedness of the component without making the inductance any greater.

Did You Know?

In some old radio receivers, you'll find air-core inductors wound around small waxed cardboard cylinders resembling short lengths of drinking straw. Some hobbyists even use waxed wooden dowels to support "air core" coils! Neither of these materials significantly increases the inductance.

Figure 3-16 shows the symbol for a tapped air-core inductor; in this case the coil has two tap points along its length. Whereas a fixed inductor has only two leads or terminals (one at either end), a tapped inductor has three or

FIG. 3-15 *Symbol for an air-wound (or air-core) inductor.*

FIG. 3-16   *Symbol for an air-core inductor with two fixed taps.*

more. When you want to tap a coil, you attach conductors to turns at intermediate points. You get maximum inductance by connecting the end leads or terminals to the external circuit. A tapped arrangement lets you select one or more pairs of points having less inductance than the full coil has.

As an alternative to taps, a coil might have a sliding contact that you can move along the entire length of the windings. The sliding contact, which connects directly to one of the end contacts by means of a shorting wire, lets you vary the inductance almost continuously (actually it happens in little jumps as you slide the contact along, one turn at a time). You can portray this type of variable inductor with either of the symbols shown in Figs. 3-17A or B.

In equipment designed for high-power RF operation, you have an alternative to the sliding-tap method of varying the inductance of an air-core coil. The coil, consisting of solid bare wire, goes around a hollow ceramic cylindrical form, and a shaft attaches to a ceramic disk (or set of disks) inside the cylinder so that you can rotate the coil and the form together. A small wheel-like contact, resembling an automobile tire rim without the tire, travels along the length of the coil as it rotates, allowing smooth, continuous adjustment of the inductance between the "wheel" and either end. Such a component is called a *roller inductor*. You'll often encounter roller inductors in radio antenna tuners and matching networks. Figure 3-18 shows a common rendition of its symbol.

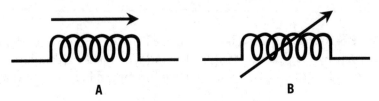

A   **B**

FIG. 3-17   *Symbols for a continuously variable air-core inductor. At A, arrow above coil symbol; at B, arrow passing diagonally through coil symbol.*

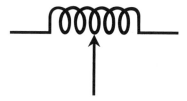

FIG. 3-18 *Symbol for a three-terminal roller inductor.*

Tip
The roller inductor symbol can also represent a sliding-contact variable inductor when the sliding contact doesn't connect to either end of the coil.

An inductor for standard AC applications, such as a 60-Hz *choke* for use in power-supply filters, can comprise a length of insulated or enameled wire wound around a solid or laminated (layered) iron core. The iron, a *ferromagnetic material*, replaces the air core. The ferromagnetic core increases the *magnetic flux density* inside the coil windings, making the inductance thousands of times greater than the inductance of an air-core coil having the same physical dimensions. Figure 3-19 shows the standard schematic symbol for a fixed-value inductor with a solid- or laminated-iron core. It's the basic coil symbol discussed earlier, along with two parallel straight lines that run for its entire length. Now and then, you'll see an iron-core inductor rendered as shown in Fig. 3-20, with the straight lines inside the coil turns.

FIG. 3-19 *Symbol for an inductor with a solid- or laminated-iron core.*

FIG. 3-20 *Alternate symbol for an inductor with a solid- or laminated-iron core.*

Some iron-core inductors have taps for sampling different inductance values. Once in awhile you'll encounter an iron-core inductor whose value you can continuously vary by pushing and pulling the core in and out of the coil. The equivalent schematic symbols for these types of inductors appear in Figs. 3-21A and B.

At high frequencies, solid-iron and laminated-iron cores aren't efficient enough to function in inductors. Engineers would say that they have too much *loss*. At frequencies above a few kilohertz (kHz), you'll need a special ferromagnetic core material if you want to increase the inductance over what you can get with *nonferromagnetic* core materials such as air, plastic, ceramic, or wood. The most common substance for this purpose consists of iron shattered into microscopic fragments, each of which has a layer of sticky, glue-like insulation applied to it. After the fragmentation and insulation process has been completed, the particles get compressed into a "solid" object called a *powdered-iron core*. Figure 3-22 shows the symbols for three different types of powdered-iron-core inductors.

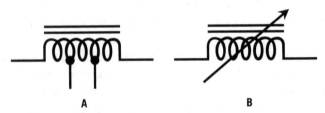

FIG. 3-21 *Symbols for a tapped coil (A) and an adjustable coil (B) with solid- or laminated-iron cores.*

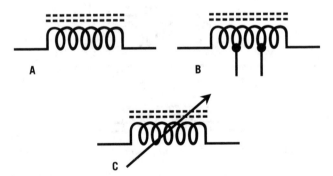

FIG. 3-22 *Symbols for fixed (A), tapped (B), and continuously adjustable (C) inductors with powdered-iron cores.*

Tip

The symbols for powdered-iron-core inductors look similar to those for solid- or laminated-iron-core inductors, except that the straight lines are broken (dashed) rather than solid. These components, like all other types of inductors, can be tapped or continuously variable.

A *transformer* contains two or more coils with the turns interspersed or wound around different parts of a single core. Figure 3-23 shows the symbol for an air-core transformer. It looks like two air-core coil symbols drawn back-to-back. Figure 3-24 shows some transformers that have iron cores. The ones at A and B have solid- or laminated-iron cores; the ones at C and D have powdered-iron cores.

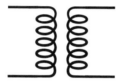

FIG. 3-23   *Symbol for a transformer with an air core.*

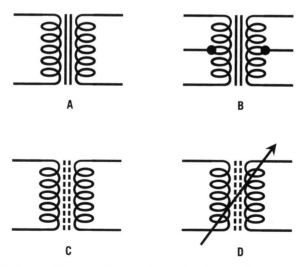

FIG. 3-24   *At A, symbol for a transformer with a solid- or laminated-iron core. At B, symbol for a transformer with a solid- or laminated-iron core and tapped windings. At C, symbol for a transformer with a powdered-iron core. At D, symbol for an adjustable transformer with a powdered-iron core.*

In a transformer, the output voltage might equal the input voltage, but often the voltages differ. In a *step-up transformer*, the output voltage is greater than the input voltage. In a *step-down transformer*, the output voltage is smaller than the input voltage. You'll find schematic symbols for these transformer types in Appendix A.

Switches and Relays

A *switch* is a component that can complete or interrupt one or more current paths. Figure 3-25 shows the symbol for a *single-pole/single-throw* (SPST) switch. It can make or break a contact at only one point in a circuit; it's a two-position device (on-off or make-break). With the switch "on" or "closed," current flows. With the switch "off" or "open," current does not flow.

Figure 3-26 is the symbol for a *single-pole/double-throw* (SPDT) switch. The *pole* coincides with the point of contact at the base of the arrowed line. The *throw* is the contact to which the arrow points. You can connect the pole to the upper throw or the lower throw, but not to both at once.

Some switches have two or more poles. In Fig. 3-27, drawing A shows the symbol for a *double-pole/single-throw* (DPST) switch, and drawing B shows the symbol for a *double-pole/double-throw* (DPDT) switch. Some switches have even more elements. The one shown in Fig. 3-28 has five poles. Engineers might call it a *five-pole/two-throw* (5P2T) switch.

> **Note!**
> In all the cases shown by Figs. 3-27 and 3-28, you must switch the poles all at once. In other words, you can't change the position of one pole without changing all the others, too.

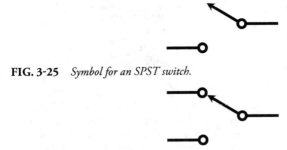

FIG. 3-25   *Symbol for an SPST switch.*

FIG. 3-26   *Symbol for an SPDT switch.*

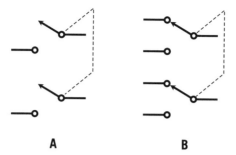

FIG. 3-27 *At A, symbol for a DPST switch. At B, symbol for a DPDT switch.*

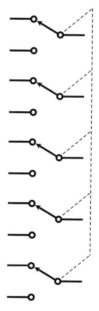

FIG. 3-28 *Symbol for a five-pole double-throw switch.*

The 5P2T device is a *multi-contact switch*. This category includes most switches that have at least two throws. For instance, a rotary switch might have a single pole and ten throw positions; Fig. 3-29 shows such a scenario. You can call this thing a *one-pole/ten-throw* (1P10T) switch!

Occasionally, you'll encounter sets of rotary switches ganged together, much like two or more variable capacitors can rotate in sync with one another. Figure 3-30 is the symbol for a ganged pair of rotary switches. The dashed line tells you that the switches mimic each other's operations. The arrowed lines indicate the throw positions, which go around in sync with each other. When

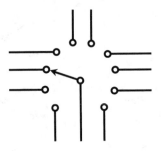

FIG. 3-29   *Symbol for a rotary switch with a single pole and ten throws.*

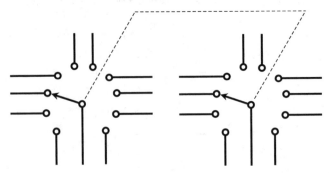

FIG. 3-30   *Symbol for a pair of ganged rotary switches, each of which has a single pole and ten throws.*

the left-hand switch pole rests at, say, throw 3 (as shown here), the right-hand switch pole also rests at throw 3.

Did You Know?

Some amateur radio operators use a unique sort of switch called a *Morse code key*. This old-fashioned device, also called a *hand key* or a *straight key*, makes or breaks a circuit so that you can send Morse code manually. In effect, it's a big SPST switch comprising a lever with a spring. The spring ensures that the key returns to the open position whenever you let go of the lever. Figure 3-31 shows its schematic symbol.

You can't always locate a switch near the circuit or system that it affects. Imagine that you want to switch a radio transmitter/receiver (or *transceiver*) between two different antennas from a control point 50 meters away. The

FIG. 3-31 *Symbol for a Morse code key.*

antennas get their signals through *coaxial cables* that carry RF current, which must remain confined to the cables if you want the system to work properly. If the cable branch point lies far away from the place where you want to put the control switch, you can use a *relay* that employs an *electromagnet* to allow remote-control switching. You install the relay at the cable branch point. You can run a length of "lamp cord," which carries plain DC, from the relay's electromagnet to your control switch.

Figure 3-32A is a functional drawing of an SPDT relay, and Fig. 3-32B shows its schematic symbol. A "springy strip" holds a movable lever, called the *armature*, to one side (which would be all the way up in this case) when no current flows through the electromagnet coil. Under these conditions, terminal X connects to terminal Y but not to Z. When sufficient DC flows in the coil, the armature moves to the other side (which would be all the way down in this case), connecting X to Z rather than to Y.

A *normally closed relay* completes a circuit when the electromagnet coil doesn't carry current, and breaks the circuit when coil current flows. ("Normal" in this sense means "no current in the coil.") In contrast, a *normally open relay* breaks the circuit when the coil doesn't carry current, and completes the circuit when coil current flows. The relay portrayed in Fig. 3-32 can serve as a normally open or normally closed relay, depending on which contacts you select. The device can also switch a single line between two different circuits.

Variations on a Theme
Figure 3-32 shows a relay in which the switch is an SPDT lever. Simpler relays have SPST switching levers, and more complicated ones have multiple-pole, single-or-double-throw lever sets.

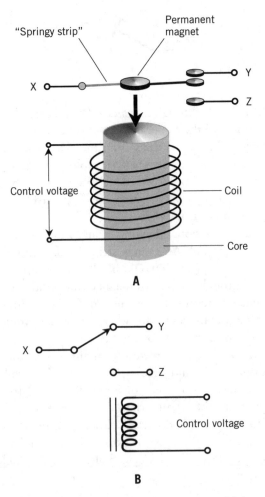

FIG. 3-32   *At A, functional drawing of an SPDT relay. At B, its schematic symbol.*

Conductors and Cables

In a schematic, a solid line commonly symbolizes an electrical conductor. Most circuits contain many conductors. When you draw a schematic of a complicated circuit, you'll often need to draw lines that cross over each other on your screen or on paper, whether the wires make contact in the real world or not.

Figure 3-33 shows two conductors that cross each other in a diagram, but that don't connect in the actual circuit. This drawing geometry doesn't neces-

FIG. 3-33 *Symbol for conductors that cross paths in a schematic but don't connect in the real world.*

sarily mean that when you build the real circuit, the conductors come near each other in that vicinity. But when you compose the schematic, you must draw one conductor across the other to avoid confusion and minimize clutter.

Aha!

A real-world circuit exists in three-dimensional (3D) space, but when you diagram it, you have to do so on a two-dimensional (2D) surface. You must learn a few tricks to ensure that your readers see things right! Someday, computerized technical manuals will provide holographic 3D "hyper-schematics" that fill a room in which you can stroll around (maybe even passing through some symbols like a ghost); but for now, such things remain the stuff of dreams.

Figure 3-34 shows two ways of portraying a point where two wires cross and *they do* electrically connect there. In the rendition at A, you "split a conductor" so it seems to contact the other one at two different points. This geometry makes it clear that the two conductors (the "split" vertical one and the "solid" horizontal one) connect in the real circuit. Black dots portray

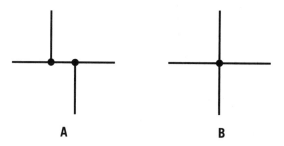

A   B

FIG. 3-34 *At A, preferred symbol for conductors that cross paths in a schematic and actually connect in the real world. At B, alternative symbol for the same scenario.*

electrical contact. In the drawing at B, the two conductors simply cross each other, and you draw a single black dot at the junction. The dot tells your readers that the conductors connect where they cross. The method shown at B might look "cleaner" at first glance, but with this neatness comes a problem: Some readers might overlook the black dot and think that the two conductors *do not* connect. The method at A avoids such misunderstandings.

Whoops!

Just as a reader might miss a black dot at a crossing point as in Fig. 3-34B, another reader might see Fig. 3-33 and imagine a black dot even though it isn't there! Then she or he will think that the two wires connect when in fact they don't. Stay away from this sort of ambiguity when you draw schematics.

In some older schematics you'll see crossed wires shown as in Fig. 3-35. These wires don't connect in the real circuit. One of the lines has a half loop or "jog" that makes it seem to "jump" over the other line. That trick (which should never have gone out of style, in my opinion) eliminates all doubt as to whether or not the actual wires connect where the lines cross.

A *cable* has one or more conductors inside a single insulating jacket. In many cases, *unshielded cables* are not specifically indicated in a schematic drawing, but appear as two or more lines that run parallel to indicate multiple conductors. *Shielded cables* require additional symbology along with the conductors. Figures 3-36 A and B show symbols for shielded cable. You'll see these symbols drawn to indicate *coaxial cable*, which comprises a single wire called the *center conductor* surrounded by a cylindrical, conduit-like conductive *shield*. At A, the shield does not connect to anything in particular, but at B, the shield connects to an *earth ground*. An insulating layer, called the *dielectric*, keeps the center conductor isolated from the shield. In most coaxial cables, the dielectric material consists of solid or foamed polyethylene.

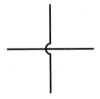

FIG. 3-35 *Archaic (but clear) representation of conductors that cross paths in a schematic but don't connect in the real world.*

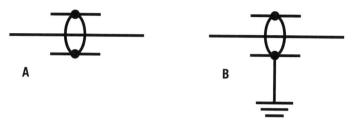

FIG. 3-36 *At A, symbol for coaxial cable with an ungrounded shield. At B, symbol for coaxial cable with an earth-grounded shield.*

Tip

Figure 3-37 shows a symbol for coaxial cable when the shield connects to a *chassis ground*, such as a metal plate on which you can build an electronic circuit. The chassis ground might go to an earth ground, but not always. In a truck, for example, no earth ground exists, so the chassis of the trucker's two-way radio must connect to the vehicle frame, the "next best thing" to an earth ground.

In some cables, a single shield surrounds two or more conductors. Figure 3-38 shows the symbol for two-conductor shielded cable whose shield goes to a chassis ground. This symbol looks like the one for single-conductor coaxial cable, except that it has an extra inner conductor. If shielded cable has more than two inner conductors, then the number of straight, parallel lines going through the elliptical part of the symbol tells you how many conductors run inside the shield. For example, if the cable in Fig. 3-38 had five inner conductors, then five horizontal lines would pass through the elliptical part of the symbol.

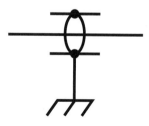

FIG. 3-37 *Symbol for coaxial cable with a chassis-grounded shield.*

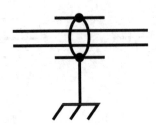

FIG. 3-38 *Symbol for two-conductor cable with a chassis-grounded shield.*

Diodes and Transistors

Figure 3-39 is the common symbol for a *semiconductor diode*. An arrow and a vertical line indicate parts of the diode, and the horizontal lines to the left and right indicate the leads. The arrowed part of the symbol corresponds to the *anode*, and the short, straight line at the arrow's tip corresponds to the *cathode*.

An *ideal diode* conducts current when electrons move *against* the arrow so the anode has a positive voltage with respect to the cathode. Engineers call that condition *forward bias*. The ideal diode does not conduct when the cathode has a positive voltage with respect to the anode. Engineers call that a state of *reverse bias*. But of course, nothing is "ideal" in this imperfect universe.

> **Whoa!**
>
> In a real-world semiconductor diode, you must apply a certain minimum forward bias, called the *forward breakover voltage*, to cause electrical conduction. Also, if the reverse bias gets too great and exceeds the so-called *avalanche voltage*, the component will conduct as if forward-biased. The exact values of the forward breakover and avalanche voltages vary from one type of diode to another. The forward breakover voltage usually falls between 0.3 V and 1.5 V; the avalanche voltage can range from a few volts to several thousand volts.

FIG. 3-39 *Symbol for a general-purpose semiconductor diode.*

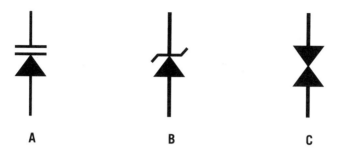

FIG. 3-40 *Symbols for a varactor diode (A), a Zener diode (B), and a Gunn diode (C).*

Figure 3-40 portrays three specialized diode types. Drawing A shows the symbol for a *varactor diode*, which can act as a variable capacitor when you apply a fluctuating reverse-bias voltage. Drawing B shows the symbol for a *Zener diode*, which can serve as a voltage regulator in a power supply that converts AC to DC. Drawing C shows the symbol for a *Gunn diode*, which can generate or amplify radio signals at extremely high and microwave frequencies.

A *silicon-controlled rectifier* (SCR) is, in effect, a semiconductor diode with an extra element and terminal. You'll see its symbol in Fig. 3-41. In the SCR representation, a circle or ellipse often (but not always) surrounds the diode symbol, and the control element, called the *gate*, appears as a diagonal line that runs outward from the tip of the arrow. In all cases, the arrow denotes the anode, and the vertical line at the arrow's tip denotes the cathode.

Figure 3-42 shows schematic symbols for *bipolar transistors*. A so-called *PNP transistor* appears at A, and an *NPN transistor* appears at B. In the PNP symbol, the arrow points away from the emitter and toward the base. In the NPN symbol, the arrow points away from the base and toward the emitter. Some engineers leave out the circle that surrounds the combined base, emitter, and collector symbols.

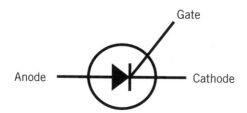

FIG. 3-41 *Symbol for a silicon-controlled rectifier (SCR).*

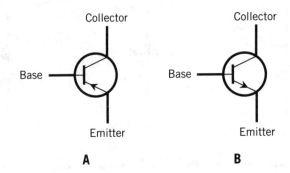

FIG. 3-42   *Symbols for a PNP bipolar transistor (A) and an NPN bipolar transistor (B).*

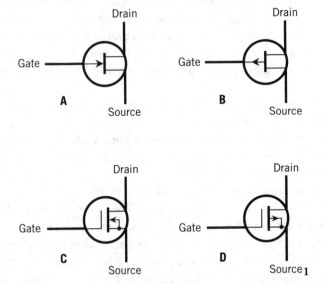

FIG. 3-43   *At A, symbol for an N-channel JFET. At B, symbol for a P-channel JFET. At C, symbol for an N-channel depletion-mode MOSFET. At D, symbol for a P-channel depletion-mode MOSFET.*

Along with the bipolar variety, you'll encounter other types of transistors. Figure 3-43 shows the symbols for four of these devices, as follows:

- At A, you see an *N-channel junction field-effect transistor* (JFET).
- At B, you see a *P-channel JFET.*
- At C, you see an *N-channel depletion-mode metal-oxide-semiconductor field-effect transistor* (MOSFET).
- At D, you see a *P-channel depletion-mode MOSFET.*

A "Current Hose"

In a field-effect transistor, the *channel* is a current path that goes directly from the source to the drain. Think of the channel as a garden hose that you can constrict to a certain extent by stepping on it. The force with which you "step on the hose" depends on the bias voltage that you apply between the gate and the source.

Tip

Manufacturers produce bipolar and field-effect transistors from various semiconductor materials and metal-oxide compounds, but the symbol, all by itself, says nothing about the chemical nature of the material. The symbol tells you the component's function, but nothing more.

A depletion-mode MOSFET has an open (conducting) channel when you don't apply any bias voltage between the source and the gate; when you do impose a bias voltage, the channel constricts and eventually closes off altogether. Then you have a state of *pinchoff*. Sometimes you'll see another type of MOSFET in electronic circuits: the *enhancement-mode* MOSFET. An enhancement-mode device has a pinched-off channel unless you apply a bias voltage between the source and the gate. Then the channel opens wider and wider as you increase the bias voltage. Figure 3-44A is the symbol for an *N-channel enhancement-mode MOSFET*. Figure 3-44B shows the symbol for a *P-channel enhancement-mode MOSFET*.

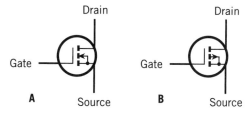

FIG. 3-44 *At A, symbol for an N-channel enhancement-mode MOSFET. At B, symbol for a P-channel enhancement-mode MOSFET.*

Operational Amplifiers

An *operational amplifier* or *op amp* is a specialized *integrated circuit* (IC) that comprises bipolar transistors, resistors, diodes, and/or capacitors, all connected together to produce or modify a signal. (Myriad types of ICs exist besides the op amp. Figure 3-45 shows the general symbol for an IC.) Sometimes you'll find two or more op amps in a single IC package; for example, you might encounter a *dual op amp* or a *quad op amp.*

Figure 3-46 is the schematic symbol for an op amp. The device has two inputs, one *non-inverting*, indicated by a plus (+) sign, and the other *inverting*, as shown by the minus (−) sign. When a signal enters the non-inverting input, the output wave emerges in *phase coincidence* ("right-side-up") with respect to the the input wave. When a signal enters the inverting input, the output wave appears in *phase opposition* ("upside-down") with respect to the input wave. The device has two power-supply connections, one for the emitters of the internal bipolar transistors (V_{ee}) and one for the collectors (V_{cc}).

When a signal comes into either input for amplification, you can place a resistor between the output and the inverting input to cause *negative feedback* that reduces or controls the gain. As you reduce the value of the resistor, the gain decreases because the negative feedback increases. This state of affairs is called the *closed-loop configuration.*

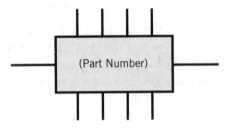

FIG. 3-45 *Generic symbol for an integrated circuit (IC).*

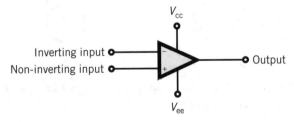

FIG. 3-46 *Symbol for an operational amplifier (op amp).*

Tip

If you don't place a resistor between the output and the inverting input, an op amp will run in the so-called *open-loop configuration*. In that case, when you apply a signal to either input, the device will produce its maximum possible gain, on the order of many thousands of times in terms of voltage! With that much gain you'll get an unstable circuit that can break into uncontrollable oscillation.

If you install a *resistance-capacitance* (RC) combination in the inverting-feedback loop of an op amp, the gain depends on the frequency of the signal that enters the device. Using specific values of resistance and capacitance, you can make a frequency-sensitive filter that provides any of four different characteristics as shown in Fig. 3-47:

• A *lowpass response* that favors low frequencies (A).
• A *highpass response* that favors high frequencies (B).
• A *resonant peak* with maximum gain at a single frequency (C).
• A *resonant notch* with minimum gain at a single frequency (D).

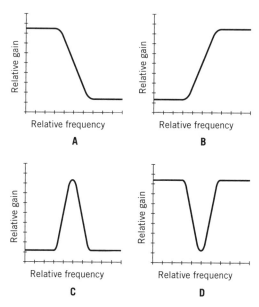

FIG. 3-47   *Gain-versus-frequency response curves. At A, lowpass; at B, highpass; at C, resonant peak; at D, resonant notch.*

Electron Tubes

Although you won't encounter *electron tubes* (often simply called *tubes*) as frequently as you would have a few decades ago, plenty of circuits and systems still use them. When you want to create the symbol for a tube, you should draw a circle and then add the necessary symbols inside the circle to portray the type of tube involved. Figure 3-48 shows the symbols for various internal tube elements.

A Fine Point

In Fig. 3-48J, the small dot indicates that the tube doesn't contain a full vacuum, but instead has rarefied gas inside. This constitutes an example of an electron tube that isn't actually a vacuum tube. All vacuum tubes are electron tubes, but the converse isn't always true!

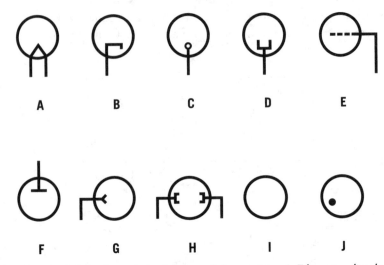

FIG. 3-48 *Symbols for electron-tube elements and characteristics. A: Filament or directly heated cathode. B: Indirectly heated cathode. C: Cold cathode. D: Photocathode. E: Grid. F: Anode (plate). G: Beam-deflection plate. H: Beam-focusing plates. I: Envelope (enclosure) for a vacuum tube. J: Envelope for a gas-filled tube.*

Figure 3-49 shows the schematic symbol for a *diode vacuum tube*. This two-element component contains an *anode* (also called a *plate*) and a *cathode*. As with a semiconductor diode, the anode normally carries a more positive voltage than the cathode when the device conducts current. When the anode has a more negative voltage than the cathode, the device generally does not conduct. The cathode emits electrons that travel through the vacuum to the anode. A hot-wire *filament*, something like a miniature incandescent bulb, heats the cathode to help drive electrons from it. In Fig. 3-49, the filament has been omitted for simplicity, a common practice in vacuum tube symbols when the filament and cathode are physically separate, an arrangement known as an *indirectly heated cathode*.

Tip

In standard symbology, all tube elements appear within a circle or ellipse, which represents the tube envelope. Occasionally (but rarely), you'll find a schematic in which a tube symbol doesn't show the envelope.

Figure 3-50 shows two versions of a *triode* vacuum tube, which consists of the same elements as the diode previously discussed, with the addition of a dashed line to indicate the grid. But another difference exists in Fig. 3-50A

FIG. 3-49 *Schematic symbol for a diode vacuum tube with an indirectly heated cathode. Although a filament exists, it's often omitted to reduce clutter.*

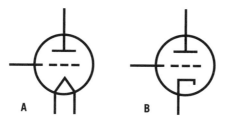

FIG. 3-50 *Symbols for triode tubes with a directly heated cathode (A) and an indirectly heated cathode (B).*

compared with Fig. 3-49. Look closely at the cathode. The tube shown in Fig. 3-50A has a *directly heated cathode,* in which the filament and the cathode are the same physical object! You apply the negative cathode voltage directly to the filament. Figure 3-50B shows the symbol for a triode tube with an indirectly heated cathode. In this case the filament resides inside the cathode, which physically takes the form of a metal cylinder running along the central vertical axis of the tube, surrounding the filament.

Tetrode vacuum tubes have two grids, so you must draw an additional dashed line as shown in Figs. 3-51 A and B. In the tetrode, the upper grid, closer to the anode, is called the *screen grid* (or simply the *screen*). Figure 3-52 shows symbols for a *pentode* tube, which has three grids and a total of five elements. In the pentode, the middle grid is the screen, and the top grid is called the *suppressor grid* (or simply the *suppressor*).

Figures 3-51A and 3-52A portray tubes with directly heated cathodes, while Figs. 3-51B and 3-52B symbolize tubes with indirectly heated cathodes.

Follow the Flow

In all the tube symbols shown here, electrons normally flow from the bottom up. They come off the cathode, travel through the grid or grids (if any), and end up at the plate. Once in awhile you'll see a tube symbol drawn sideways. In any case, the electrons go from the cathode to the plate under normal operating conditions.

Some tubes consist of two separate, independent sets of electrodes housed in a single envelope. These components are called *dual tubes.* If the two sets of electrodes are identical, the entire component is called a *dual diode, dual triode, dual tetrode,* or *dual pentode.* Figure 3-53 shows the schematic symbol for a dual triode vacuum tube with indirectly heated cathodes.

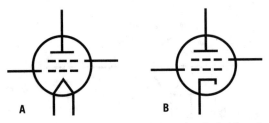

FIG. 3-51   *Symbols for tetrode tubes with a directly heated cathode (A) and an indirectly heated cathode (B).*

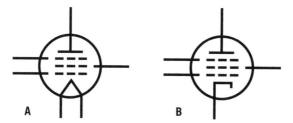

FIG. 3-52 *Symbols for pentode tubes with a directly heated cathode (A) and an indirectly heated cathode (B).*

FIG. 3-53 *Symbol for a dual triode tube with indirectly heated cathodes.*

In old radio and television receivers, you'll sometimes find tubes with four or five grids. These tubes have six and seven elements, respectively, and are called a *hexode* and a *heptode*. They work for *mixing*, a process in which two RF signals having different frequencies combine to produce new signals at the sum and difference frequencies. Figure 3-54A shows the schematic symbol for a hexode; Fig. 3-54B shows the symbol for a heptode. Some engineers call the heptode a *pentagrid converter*.

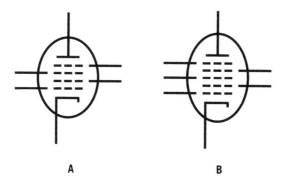

FIG. 3-54 *At A, symbol for a hexode tube. At B, symbol for a heptode tube, also called a pentagrid converter. Both symbols show tubes with indirectly heated cathodes.*

Tip

You won't encounter hexodes and heptodes in modern electronics, but if you like to work with antique radios, you should get familiar with them. However, take this warning: You'll probably have a difficult time finding a replacement component, should one of these relics go "soft" on you!

Electrochemical Cells and Batteries

You'll often see a *cell* or *battery* employed as the power source for a circuit or system. Figure 3-55 shows the schematic symbol for a single *electrochemical cell*, such as the sort that you'll find in a flashlight. Most such cells produce approximately 1.5 V DC.

Electrochemical batteries with higher voltage outputs comprise multiple cells connected in series (negative-to-positive in a chain or string); the symbol for a multicell battery takes this design into account as shown in Fig. 3-56.

If a circuit calls for a multicell battery in the form of two or more discrete single cells in series, you can draw the single-cell symbols individually with wire conductor symbols between them. Figure 3-57 shows an example with three cells.

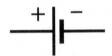

FIG. 3-55 *Symbol for a single electrochemical cell.*

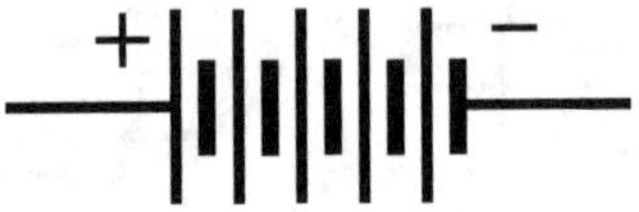

FIG. 3-56 *Symmbol for a self-contained multicell electrochemical battery.*

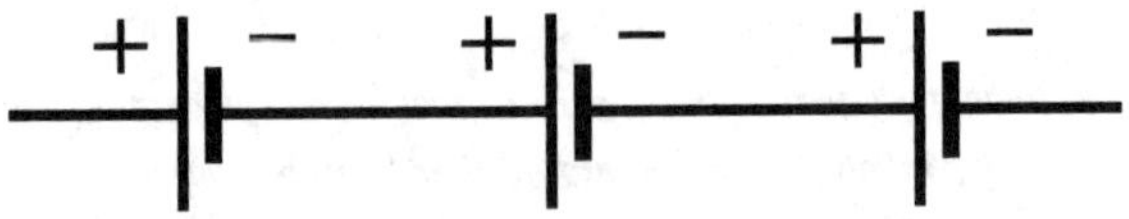

FIG. 3-57 *Symbol for three single electrochemical cells connected in series to form a battery.*

Tip

When you place two or more individual cells in a "battery holder" designed for series connection, you can draw the multicell battery symbol (Fig. 3-56) to denote the set of cells. You need not render the cells one by one (as in Fig. 3-57).

Standard practice calls for polarity signs to go with the symbols for cells or batteries. Unfortunately, some people neglect this detail. Then when you look at the schematic, you'll have to infer the cell or battery polarity by scrutinizing the rest of the circuit. (The longer end line usually goes with the positive terminal, but not always.)

Logic Gates

All digital electronic devices employ switches that perform specific logical operations. These switches, called *logic gates*, can have from one to several inputs and a single output. Logic devices have two states, represented by the digits 0 and 1. In most cases, 0 represents the *low state* and 1 represents the *high state*.

- A *logical inverter*, also called a *NOT gate*, has one input and one output. It reverses, or inverts, the state of the input. If the input equals 1, then the output equals 0. If the input equals 0, then the output equals 1. Figure 3-58A shows its symbol.
- An *OR gate* can have two or more inputs (although it usually has only two). If both, or all, of the inputs equal 0, then the output equals 0. If any input equals 1, then the output equals 1. Mathematicians would say that such a gate performs an *inclusive-OR operation*. Figure 3-58B shows its symbol.
- An *AND gate* can have two or more inputs (although it usually has only two). If both, or all, of the inputs equal 1, then the output equals 1. If any input equals 0, then the output equals 0. Figure 3-58C shows its symbol.
- An OR gate can be followed by a NOT gate. This combination gives you a *NOT-OR gate*, more often called a *NOR gate*. If both, or all, of the inputs equal 0, then the output equals 1. If any inputs equals 1, then the output equals 0. Figure 3-58D shows its symbol.

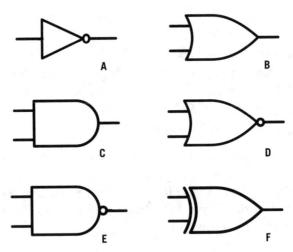

FIG. 3-58   *Symbols for a logical inverter or NOT gate (A), an OR gate (B), an AND gate (C), a NOR gate (D), a NAND gate (E), and an XOR gate (F).*

- An AND gate can be followed by a NOT gate. This combination gives you a *NOT-AND gate*, more often called a *NAND gate*. If both, or all, of the inputs equal 1, then the output equals 0. If any input equals 0, then the output equals 1. Figure 3-58E shows its symbol.
- An *exclusive OR* gate, also called an *XOR gate*, has two inputs and one output. If the two inputs have the same state (both 1 or both 0), then the output equals 0. If the two inputs have different states, then the output equals 1. Mathematicians would say that such a gate performs the *exclusive-OR operation*. Figure 3-58F shows its symbol.

Here's a Quickie!
Table 3-1 denotes the functions of the six common digital logic gates in abbreviated form for quick reference.

Summary

This chapter identifies most of the symbols that you'll see in schematics, but plenty of symbols exist that are not included here. You'll find a comprehensive table of schematic symbols in Appendix A. You'll see symbols for jacks and plugs, piezoelectric crystals, lamps, microphones, meters, antennas, and many other components.

TABLE 3-1 *Logic gates and their characteristics.*

Gate type	Number of inputs	Remarks
NOT	1	Changes state of input.
OR	2 or more	Output high when any inputs are high.
		Output low when all inputs are low.
AND	2 or more	Output low when any inputs are low.
		Output high when all inputs are high.
NOR	2 or more	Output low when any inputs are high.
		Output high when all inputs are low.
NAND	2 or more	Output high when any inputs are low.
		Output low when all inputs are high.
XOR	2	Output high when inputs differ.
		Output low when inputs are the same.

Do you expect to have a difficult time memorizing all these symbols? Well, don't worry; you'll get to know them over time. Look at some simple schematics after you've read this chapter. Refer to Appendix A whenever you see a symbol that you can't identify. Within a few hours you'll want to move on to more complex schematics, again referencing the unknown symbols. After a few weekends of practice, you'll recognize most symbols without even thinking.

In electricity and electronics, most symbols derive from the structures of the components they represent. Schematic symbols often appear in groups, each of which bears some relationship to the others. For example, you'll encounter many different types of transistors, but they all look somewhat alike. The same rule applies to the symbols for diodes, resistors, capacitors, inductors, transformers, meters, lamps, and most other electronic components. *Most*, but not *all*. A few "renegade" symbols exist that defy reason. All you can do with these things is memorize them, scratch your head, and laugh.

4

Simple Circuits

Among people who teach about schematics, two schools of thought prevail. Some mentors say that you should learn to read schematics *before* you learn to draw them. Others insist that you should learn to read them *while* you learn to draw them. Either view has its merits, so let's take advantage of them both! You can start by learning (okay, memorizing) the basic component symbols. Then you can try to read all the schematics that you see. When this business begins to bore you, then you can draw your own schematics.

> **Tip**
> If you alternately read and draw schematics, you'll get comfortable with them before long. I suggest that you devote half of your study time to reading diagrams, and the other half of your time to drawing them.

Getting Started

This chapter deals with simple circuit diagrams in pictorial and schematic form. You can cobble a circuit together and then draw a schematic of the "beast" that you've created. Once in awhile, this approach works on the first try, but such good luck doesn't strike very often. Most engineers and inventors draw a schematic to start, and then build and test the actual circuit on the basis of the diagram.

If the design is new or experimental, some bugs (flaws) will probably exist in the first version, called the *prototype*, necessitating various component deletions, additions, or substitutions. Every time you modify the test circuit, you can note the results and change the schematic accordingly. In the end, the finished and corrected schematic will reflect the result of design theory, real-time testing, and "tweaking."

Figure 4-1 shows a simple circuit that you've likely built at one time or another. Basically, it's a flashlight without the external case and switch. The device consists of a single electrochemical cell and an incandescent light bulb. This pictorial also shows the conductors, which attach to the bulb and the cell. The conductors carry current out of the cell, through the bulb, back through the cell, through the bulb again, around and around until the cell dies or the bulb burns out.

Follow the Flow

In the circuit of Fig. 4-1, electrons travel from the negative terminal of the cell through the bulb and back to the positive terminal of the cell, "leapfrogging" from atom to atom in the metal wire and the bulb's *filament*. That's how most electricians and engineers look at this situation. But many physicists will tell you that the current goes from the positive cell pole to the negative cell pole! That's called *theoretical current* or *conventional current*. As an engineer, you'll find it easier to imagine current flowing in the same direction as the electrons move: from negative to positive.

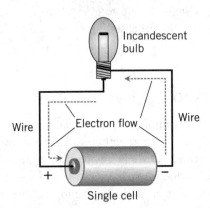

FIG. 4-1 *Pictorial drawing of a flashlight circuit comprising an electrochemical cell, some wire, and an incandescent bulb.*

In order to draw a schematic of the flashlight illustrated in Fig. 4-1, you must know three symbols. They represent the cell, the conductors, and the bulb (Fig. 4-2). Once you know the symbols, you can draw a schematic based on the appearance of the circuit in the pictorial.

That's Weird!

In most physical flashlight cells, the positive (+) end looks like a small bump or button, while the negative (−) end is large and flat, occupying almost the entire cross section of the cell. In the schematic symbol of the same cell, the positive end has a long line and the negative end has a short line, the opposite of what you'd expect.

Start by drawing the cell symbol. You can think of the cell as the "heart" of the circuit because it "pumps" electrons through everything. Next comes the symbol for the bulb, which you can draw at any point near the cell symbol. Using this example, you should try to arrange the schematic symbols in the same relative positions as they appear in the pictorial, so you'd place the bulb symbol above the cell symbol.

Now that you've drawn the two major symbols, you can use conductor symbols (solid black lines) to connect them together. Notice that the pictorial shows two conductors. Therefore, the schematic diagram also shows two conductors. Figure 4-3 shows the completed schematic, the symbolic equivalent of Fig. 4-1.

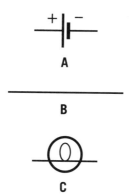

A

B

C

FIG. 4-2 *Schematic symbols for an electrochemical cell (A), an electrical conductor such as wire (B), and an incandescent bulb (C).*

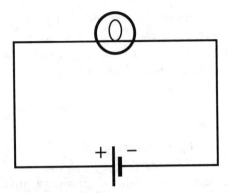

FIG. 4-3 *Schematic diagram of a single-cell flashlight.*

Tip
Figure 4-3 isn't the only way that you can portray this circuit in schematic form. But any schematic representation will need the same set of basic symbols: cell, bulb, and conductors. The only possible changes involve the positioning of the component symbols on the page.

Figure 4-4 shows two alternative schematics of the single-cell flashlight diagrammed in Fig. 4-3. All three of these diagrams (the one in Fig. 4-3 and the two in Fig. 4-4) represent the same circuit, but they look different because of the relative positions of the symbols on the page.

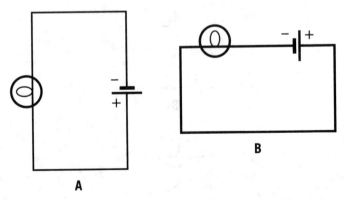

FIG. 4-4 *Alternative arrangements for the flashlight schematic. At A, cell on the right and bulb on the left; at B, cell and bulb both on the top.*

Figure 4-5 shows a flashlight with a switch and two cells. When you examine this pictorial, you'll see that any schematic rendition must have two cell symbols, the conductors, the bulb, and the switch. Figure 4-6 shows the symbols that you'll need to produce a complete schematic of this flashlight. Again, you should draw the symbols in the same sequence as the components appear in the pictorial. Figure 4-7 shows the result. The two cell symbols are drawn separately, connected in series, with polarity markings for each one. In the series connection, the positive terminal of one cell goes to the negative terminal of the other. You'll use two conductors from the cell terminals and a third conductor to connect the switch to the light bulb, and you might also need a fourth conductor connecting the cells together to form a battery (unless the cells rest directly against each other, a common arrangement in manufactured flashlights). Figure 4-7 shows the switch in the "off" position.

Eureka!

Now you know what a common two-cell flashlight looks like when represented with schematic symbols. The next time you switch one of those things on, you can imagine the switch symbol in Fig. 4-7 moving from the "off" (or open) position to the "on" (or closed) position.

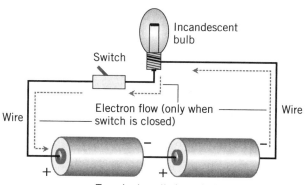

FIG. 4-5 *Pictorial of a flashlight circuit using two electrochemical cells in series, some wire, a switch, and an incandescent bulb.*

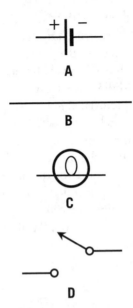

FIG. 4-6 *Symbols for components in the two-cell switched flashlight: Cell (A), wire (B), bulb (C), and switch (D).*

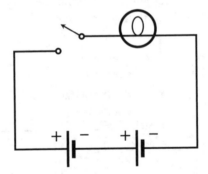

FIG. 4-7 *Schematic of the two-cell switched flashlight.*

Figure 4-8 is a pictorial drawing of an *RF field-strength meter*. Radio engineers sometimes use this type of meter to 2see whether or not an RF electromagnetic (EM) field exists at a given location. You'll find this device handy if you want to locate the source of something that's causing RF interference to your amateur or shortwave radio receiver. The circuit consists of a small whip antenna, an RF diode, a microammeter (a sensitive current meter graduated in millionths of an ampere), and an inductor.

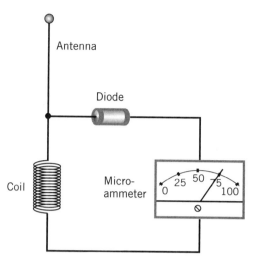

FIG. 4-8 *Pictorial of an RF field-strength meter circuit.*

To draw a schematic of this circuit, you must know the symbols for an antenna, an inductor (in this case an air-core coil), a microammeter, and a diode. Figure 4-9 shows these symbols individually. You should connect the symbols in the same sequence as you do when you follow the pictorial around.

Figure 4-10 is a schematic of the field-strength meter shown pictorially in Fig. 4-8. This drawing involves nothing more than substitution of the schematic symbols for the pictorial symbols. When you assemble the meter, the parts need not go in the same *physical locations* as the schematic implies, but you should connect them in the same *sequence* as the schematic indicates.

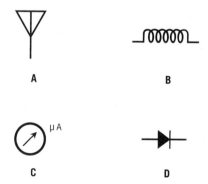

FIG. 4-9 *Symbols for the components in the field-strength meter: Antenna (A), coil (B), microammeter (C), and diode (D).*

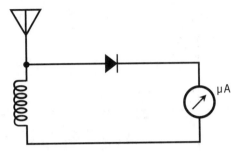

FIG. 4-10   *Schematic of the RF field-strength meter.*

Caution!

When you build a circuit from a schematic, *triple-check* your component interconnections to ensure that they agree with the schematic. If you try to build the circuit according to Fig. 4-10 and make a mistake anywhere in the wiring (and it's easy to do), you can't expect the finished device to work. In some cases, a minor wiring error can cause major component damage!

Follow the Flow

In the circuit of Figs. 4-8 and 4-10, an EM field induces RF current in the antenna and inductor. That current constitutes high-frequency AC. The diode "chops off" either the positive or negative half of every cycle (depending how you connect it) to produce *pulsating DC* like the rectifier in a power supply for home appliances, but at a higher frequency. The microammeter tells you the current level. As the EM field gets more intense, the current increases, and the meter reading goes up.

Now let's examine a more complicated circuit. Figure 4-11 is a schematic of a power supply that produces battery-like DC from utility AC. As you read the diagram from left to right, you'll see that the power plug goes to the transformer primary winding (the coil on the left-hand side of the two vertical lines) through a fuse. The top end of the transformer secondary winding (the coil on the right-hand side of the two vertical lines) connects to the anode of a rectifier diode. The diode's cathode goes to the "plus" side of an electrolytic capacitor. The "minus" side of the capacitor (not labeled because you can

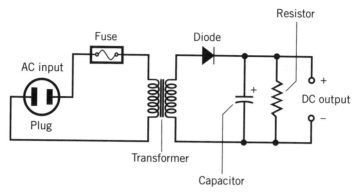

FIG. 4-11 *Schematic of a simple DC power supply.*

infer it on the basis of the "plus" sign alone) goes to the bottom end of the transformer secondary. A fixed resistor shunts (connects in parallel with) the capacitor. The DC output appears at the extreme right, across the resistor.

The physical size and weight of a real-world power supply, which you can build on the basis of Fig. 4-11, will depend on the voltage and current that you want to get from it. Because any DC power supply has a polarized output, you should use "plus" and "minus" signs to show which output terminal provides the positive voltage and which terminal provides the negative voltage. Any power supply that uses a single diode, capacitor, and resistor will have this configuration. Whether the DC output terminals provide 5 V at 1 A or 5000 V at 50 A, the schematic will look the same. But Figure 4-11 says nothing about how many volts or amperes the transformer, diode, capacitor, and resistor can handle.

Follow the Flow

In the circuit of Fig. 4-11, utility AC enters the plug on the left. The AC passes through the fuse and flows in the transformer primary. In the secondary, AC also flows, but the voltage across the secondary might be higher or lower than the voltage across the primary (depending on the transformer specifications). The diode allows current to flow in only one direction; in this case electrons can go only from right to left. As a result, pulsating DC comes out of the diode. The capacitor "filters out" the pulsations, called *ripple*, in the DC output from the diode; engineers call it a *filter capacitor*. The resistor discharges, or *bleeds*, the capacitor when you switch off the power supply and/or unplug it from the utility outlet; engineers call it a *bleeder resistor*.

Bleeding for Your Life

If the bleeder resistor is missing from Fig. 4-11 and you disconnect the power-supply output from the system for which it normally provides voltage, then the capacitor will hold a charge for a long time after you switch the supply off, even if you unplug the entire circuit from the AC utility mains. If the supply is designed to produce high DC voltage (more than about 12 V), you might receive a deadly shock from the output terminals even if the supply has been offline for hours! That's why all well-designed high-voltage power supplies include bleeder resistors.

Component Labeling

Figure 4-12 is a schematic of the same circuit as the one shown in Fig. 4-11, but in this case each component has an alphabetic/numeric designation. These designators all refer to a list at the bottom. Now you know that this power supply uses a transformer with a primary winding rated at 120 VAC and a secondary winding that yields 18 VAC. The circuit has a diode rated at 50 *peak inverse volts* (PIV) and a *peak forward current* of 1 A; a 100 *microfarad*, 50 V capacitor; and a 10,000 ohm, 1 W carbon resistor. The fuse is rated for a maximum current of 0.5 A at 120 V.

An AC Voltage Quirk

When electricians say that an outlet provides 120 VAC, they usually mean 120 *effective volts*. With AC, the *effective voltage* differs from the *peak voltage*. The most common expression for effective AC voltage is called *root-mean-square* (RMS). For most AC waves, the peak voltage equals roughly 1.414 times the RMS voltage, so the RMS voltage equals roughly 0.707 times the peak voltage.

The letters that identify each component are fairly standard in the industry. In Fig. 4-12, a numeral 1 follows each letter. The designation T1, for instance, indicates that the component is a transformer (T) and that it's the first such component referenced. If this circuit had two transformers, then one of them would bear the label T1 and the other one would bear the label

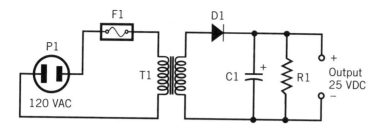

C1 – 100 microfarad electrolytic rated at 50 volts
D1 – 50 peak inverse volts at 1 ampere
F1 – 0.5 ampere at 120 volts
P1 – Male line plug
R1 – 10,000 ohm 1 watt carbon
T1 – 120 volt primary, 18 volt secondary, 1 ampere

FIG. 4-12 *Schematic of the power supply, including component designators and specifications.*

T2. The numerals reference the position or order on the components list. The diode has the reference D1, with D serving as the standard abbreviation for most diodes. Standardization is not universal, however! In some instances, the diode might bear the label SR1, where the letters SR stand for *silicon recti-fier.* Some Zener diodes have labels such as ZD1, ZD2, and so on. The labels don't matter as long as you write the component designators right next to the corresponding symbols. If you replaced the designation D1 with SR1, your readers would still know that the abbreviation went with the symbol for the diode, as long as you put the abbreviation close to the symbol.

In the situation of Fig. 4-12, you might not want to include a numeral next to each component designation, because you need only one of each component to build the circuit! You could write P for the plug, F for the fuse, T for the transformer, D for the diode, C for the capacitor, and R for the resistor. Or, if you had confidence that your readers knew all the symbols, you could leave out the designators altogether. Nevertheless, standard diagramming practice requires that you always include a letter and a numeral, even if only one of a certain component type exists.

In a complicated electronic system, you might find several hundred components of the same type (resistors, for example), many of which come from the same family. For instance, if you see the designation R101, then you know that the system contains at least 101 resistors. If you want to know the type and value of resistor R101, you will have to look up R101 in the components list to find its specifications.

> **Tip**
> You can use the schematic of Fig. 4-12 to build a power supply with an output of approximately 25 VDC if you don't expect it to deliver much current. With a *light load* (a low current demand), the filter capacitor will hold the DC voltage near the positive peak AC voltage.

Table 4-1 shows the standard letter designations for most types of electronic components that you'll see in schematic diagrams. Some of these abbreviations vary in real-world documentation, depending upon the idiosyncrasies of the person making the drawing or designing the circuit. Most designations comprise the first letter or letters of the component names. If the component has a complex name, such as *silicon-controlled rectifier*, the first letters from each word can be used, so you get SCR1. You'll designate a resistor by R, a capacitor by C, a fuse by F, and so on. Conflicts arise, of course. If you want to specify a relay, you must use some letter other than R, because R indicates a resistor. The same thing happens if you want to label a crystal; you can't use C because that letter refers to a capacitor. If you examine Table 4-1 often as you read and draw schematic diagrams, you'll absorb all its information after awhile.

The power supply of Fig. 4-13 has a *full-wave bridge* rectifier along with a better ripple filter than the lone capacitor in the circuit of Fig. 4-12. The inductor, L1, is a filter choke, which, along with capacitor C1, does an excellent job of smoothing out the DC so it closely mimics what you'd get from a battery. The Zener diode, D5, keeps the DC output voltage from exceeding

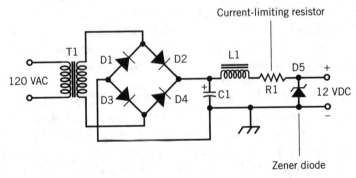

FIG. 4-13 *Schematic of a power supply that uses full-wave bridge rectification (four rectifier diodes) and Zener-diode voltage regulation.*

TABLE 4-1 *Letter abbreviations for component symbols that you'll often see in schematic diagrams.*

Abbreviation	Full Component Name
ANT	Antenna
B	Battery
C	Capacitor
CB	Circuit board
D	Diode
EP	Earphone
F	Fuse
GND	Ground
I	Incandescent lamp
IC	Integrated circuit
J	Receptacle, jack, or terminal strip
K	Relay
L	Inductor
LED	Light-emitting diode
M	Meter
NE	Neon lamp
P	Plug
PC	Photocell
PH	Earphone
Q	Transistor
R	Resistor
RFC	Radio-frequency choke
RY	Relay
S	Switch or telegraph key
SCR	Silicon-controlled rectifier
SPK, SPKR	Speaker
SR	Selenium rectifier
T	Transformer
TP	Terminal or test point
U	Integrated circuit
V	Vacuum tube
Y	Quartz crystal
Z	Circuit assembly

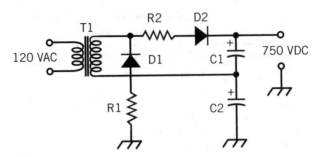

FIG. 4-14 *Schematic of a voltage-doubler power supply.*

12 V. The current-limiting resistor keeps D5 from burning out, but it also prevents you from using the power supply with any appliance that demands a lot of current. (If you ask the supply for too much current, the output voltage will drop below 12 VDC.)

Figure 4-14 shows the circuit for a *voltage-doubler power supply.* The two capacitors, C1 and C2, charge up from the full transformer secondary output after going through diodes D1 and D2. Because the two capacitors are connected in series, they behave like batteries in series, giving you twice the voltage. But there's a catch! A voltage-doubler power supply works well only at low current levels. If you demand too much current from one of these supplies, you'll "draw down" the capacitors and the voltage will drop. You'll end up with more ripple too, because C1 and C2 won't be able to smooth out the DC very well.

In Figs. 4-13 and 4-14, the letter designations are the same for each component type, but the numerals advance, one by one, up to the total number of units. So, for example, in Fig. 4-13 you see diodes D1 through D5 because the circuit contains five diodes. (The Zener diode to the right of R1 has the letter D just like the rectifier diodes have, but you can tell it's a Zener diode because of the bent line in the symbol.) All the other components have only one of each type. In Fig. 4-14, you see two diodes, two capacitors, and two resistors, so the numerals for D, C, and R go up to 2. The circuit has only one transformer, so you'll see only the numeral 1 after the letter T.

Tip
Even when multiple components have identical value (820 ohms, for example, or 50 microfarads), you must give them separate *numerical* designations when you put them all in a single circuit.

Schematics don't reveal every physical aspect of a device the way photographs or detailed pictorials can do. Schematics show you schemes, that's all! A schematic allows you to make the correct electrical connections as you assemble a circuit. It also lets you locate individual components when you test, adjust, debug, or troubleshoot the circuit.

If you find the foregoing discussion too philosophical, maybe a real-world example will clear things up. In a schematic, a solid line represents a conductor. Maybe that conductor is a wire, but maybe not! It might be a component lead or a *foil run* on a *printed circuit* (PC) board. Whether or not you need a length of wire to connect two components will depend on how close together those components reside in the physical layout. Examine the simple schematic of Fig. 4-15. The circuit contains three resistors, all of which go together in a parallel arrangement. If you take the schematic literally and describe the situation with rigor, you can recite the following facts.

- One conductor connects the left-hand side of R1 to the left-hand side of R2.
- A second conductor connects the left-hand side of R2 to the left-hand side of R3.
- A third conductor connects the right-hand side of R1 to the right-hand side of R2.
- A fourth conductor connects the right-hand side of R2 to the right-hand side of R3.
- A fifth conductor connects to the left-hand sides of all three resistors and runs off to the left, out of sight.
- A sixth conductor connects to the right-hand sides of all three resistors and runs off to the right, out of sight.

In practice, you can make some or all of these connections by attaching wires to the resistor leads, but if the components lie close enough to each other in the real world, you can use the leads themselves to make the connections. Then Fig. 4-15 will represent the physical arrangement shown pictorially in Fig. 4-16.

If you want to follow good engineering principles, you'll make all of your electronic circuits as compact (and dependable) as possible by using a minimum of point-to-point wiring and trying to make the component leads serve for interconnection purposes whenever you can. Of course, in the above example, if the three resistors had to go in different parts of the circuit separated by long distances, then you'd have to use separate interconnecting conductors between them.

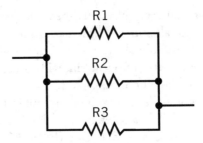

FIG. 4-15   *Schematic of three resistors connected in parallel.*

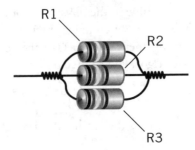

FIG. 4-16   *Pictorial of three resistors connected in parallel with their leads twisted together.*

Tip
As you design the physical layout of a circuit, you should strive to minimize the lengths of the interconnecting wires or foil runs.

Troubleshooting with Schematics

Engineers and technicians use schematics to design electronic systems, and schematics can also prove valuable for troubleshooting equipment when problems develop. Nevertheless, merely knowing how to read schematics isn't enough. You must also know what the individual components do, and how diverse circuits can work together in the complete system. No matter how proficient you get at electronics troubleshooting, seemingly simple repair jobs can morph into huge headaches without complete, accurate, and clear schematic representations of the hardware.

> **Remember!**
> Schematics clarify the interconnections among circuit components. Schematics portray circuit elements in a logical and easy-to-understand manner. However, these diagrams tell you little or nothing about the component layouts in real-world circuits, devices, and systems.

When you build a circuit on the basis of a schematic, the physical object rarely bears much physical resemblance to the diagram. You can't build a complex electronic system by placing the components in the same geometric positions as they appear in the schematic. The schematic arranges the symbols on the page for easy reading. Schematics occupy two dimensions, whereas real-world components, circuits, devices, and systems occupy three dimensions. You need only to look inside of a major electronic system such as a television set or computer to realize the complexities that you'd face in troubleshooting such a system without the help of a schematic.

If you know a fair amount about electronic components and how they operate in various circuits, then you can use a schematic to get a good idea of where a particular problem might occur. Then, by testing various circuit parameters at these critical points and comparing your findings with what the schematic diagram says should exist at those points, you can make a quick assessment of the trouble. For example, if a schematic shows a direct connection between two components in a circuit but a check with an ohmmeter reveals high resistance between the two, then you can assume that a conductor has broken or a contact has shaken loose. If a schematic shows only a capacitor between two components (with no other circuit routes around it) and a reading with your ohmmeter shows zero ohms or only a couple of ohms, then you can assume that the capacitor has shorted out and you'll have to replace it.

Beginners to electronics troubleshooting and diagram reading sometimes assume that a professional can instantly isolate a problem to the component level by looking at the schematic. This idealized state of affairs might prevail for a few simple circuits, but in complex designs the situation grows more involved. A schematic can help a technician make educated guesses as to where a problem lies or what causes it, but an exhaustive diagnosis nearly always requires testing.

You're the Sleuth!

A malfunction in an electronic device might not have a single, easy-to-identify cause. Sometimes several possible causes exist, and you must whittle the problem down to a single cause by following a process of elimination.

Suppose that a circuit won't activate, and you can't detect any voltage when you check all the test points shown in its schematic. In all probability, no current is passing through the circuit at all. However, you don't know from this observation exactly what has caused the failure. Ask yourself the following questions.

- Has one of the components in the power supply gone bad?
- Have you accidentally pulled the power cord from the wall outlet?
- Has a conductor broken between the power-supply output and the device input?
- Has the power-supply fuse blown?

In a situation like this, you'll want to consult the schematic as you go through all of the standard test procedures. You might find the test point that serves as the power supply input, indicated on the schematic. If you check the voltage at this point and it appears normal, then you can assume that the problem lies somewhere further along in the circuit. The schematic and the test instrument readings allow you to methodically search out and isolate the problem by starting at a point in the circuit where operation appears normal and proceeding forward until you get to the point where the circuit behaves abnormally.

You can test in the other direction as well. If no output comes from the power supply, you know that you must search backward toward the trouble point. You'll keep testing until you reach a point of normal operation and then check all the components in that vicinity. With the help of a schematic, you'll follow your progress and narrow the problem area down to something between two points (the point farthest back from the output at which the problem exists, and the point furthest forward from the input where things test normal). Chances are good that this narrowing process will isolate the trouble to a single component or circuit connection.

Go back and look again at the flashlight circuit of Fig. 4-7. Although the schematic doesn't say so, the two batteries in series should yield 3 VDC, because a typical flashlight cell provides 1.5 VDC all by itself, and DC voltages add up in series connections. Some schematics provide voltage test points and maximum/minimum readings that you should expect, but this simple example doesn't.

Suppose that the flashlight has stopped working, and you decide to test the circuit with a *volt-ohm-milliammeter* (VOM), also called a *multimeter*, with the help of Fig. 4-7. First of all, you can measure the individual voltages across the cells. With the meter's positive probe placed at the positive cell terminal and the negative probe at the negative terminal, you should get a reading of 1.5 V across each cell. If both cells read 0 V, then you know that both cells have lost all their electrical charge. If one cell reads normal and the other one reads 0 V, then in theory you should only have to replace the cell that appears "dead." (In practice, if one or more cells in a set appears "dead" or "dying," you should replace the whole set at the same time, even if some of them test okay). If both cells appear normal, then you can test the voltage across the bulb. Here, you should expect a reading of 3 V under normal operation with the switch closed. If you do indeed observe 3 V here, then you can diagnose the problem by looking at the schematic. The bulb must have burned out! The schematic shows you that current should pass through the bulb if it can conduct, so it must light up. But if the filament has broken, no current can flow through the bulb, so it won't light up. In fact, with a burned-out bulb, no current will flow anywhere in the circuit.

On the other hand, let's say that you see 3 V across the pair of cells, but 0 V at the light bulb. Apparently, a break exists somewhere in the circuit between the bulb and the two-cell battery. Three conductors are involved here: one between the negative terminal of the battery and one side of the bulb, another between the positive battery terminal and the switch, and another between the switch and the other side of the bulb. You can conclude that one of the conductors has broken, a contact has been lost where the conductor attaches to the battery, or the switch is defective. While you keep an eye on the schematic, you can test for a defective switch by placing the negative meter probe on the negative battery terminal and the positive probe on the input to the switch. If you see a normal voltage reading, then the switch must be defective. If you still get no voltage reading, then one of the conductors must have come loose or broken.

Admittedly, the scenario just described presents only a basic example of troubleshooting using a schematic. But imagine that the flashlight has been built according to some new design that you know nothing about. Then you'll need the schematic an adjunct to the standard test procedures with the VOM. You'll use this same procedure whenever you test complex electronic circuits of a similar nature.

> **Tip**
> In most situations, no matter how complicated a circuit looks, you can break it down into a combination of simpler circuits. If you must perform a comprehensive troubleshooting operation, you might have to test every simple circuit individually.

A More Sophisticated Diagram

Figure 4-17 shows a schematic presented in a form that can assist a troubleshooting technician. The circuit has a single NPN bipolar transistor along with some resistors and capacitors. Note that *test points* (abbreviated TP) exist at three different locations: TP1 at the emitter of the transistor, TP2 at the base, and TP3 at the collector.

If you need to troubleshoot this circuit (which happens to be a low-power amplifier of the sort you might find in a vintage radio receiver or ultrasound-actuated switch), you'll connect your VOM between chassis ground and each test point in turn. You'll note the meter readings and compare them with known normal values.

In many circuits, the actual voltages can deviate from the design values by up to 20 percent; if this information is important, you'll usually find it at the bottom of the schematic or in the accompanying literature. If you get readings that fall within this known error range (called the *component tolerance*), then you can reasonably assume that this part of the circuit is working properly.

However, if you get readings far outside of the tolerance range, then you should suspect a problem with the associated circuit portion, or possibly with other circuits that feed it.

Follow the Flow

The circuit of Fig. 4-17 receives a weak AC input signal (such as might come from an ultrasonic pickup) and boosts it so that it can drive a device that consumes significant power (such as an electronic switch). The signal flows generally from left to right. The original AC signal enters at the input terminals, passes through capacitor C2, and reaches the base (left-hand electrode) of transistor Q1. The base acts as a variable "current valve" that causes large current fluctuations through Q1 as the electrons flow from ground through R1 to the emitter (the bottom electrode), then onward to the collector (top electrode), and out through R4 to the positive power supply terminal. Capacitors C1 and C3 allow AC signals to pass while blocking DC from the power supply, so the DC won't upset the operation of external circuits. Capacitor C2 keeps the transistor's base at a constant DC voltage while allowing the input signal to enter unimpeded. Resistors R2 and R3 have values that engineers have carefully chosen to place precisely the right DC voltage, called bias, on the base of Q1, ensuring that the transistor will work as well as possible in this application.

Many schematics that accompany electronic equipment, especially "projects" that you build from kits containing individual components, include information that can help you not only in troubleshooting, but also in the testing and alignment procedures that you must follow after you've completed the assembly process but before you put the equipment into service. As a further aid, the literature might include pictorials that show you where each part belongs on the circuit board or chassis. That way, you can follow the circuit not only according to its electrical details, but also along the physical pathways as they look in real life.

According to standard schematic drawing practice, every component should bear a unique alphabetic/numeric label to designate it, as you see in Fig. 4-17. However, a few alternative labeling forms are also acceptable. Figure 4-18 shows the same circuit as the one in Fig. 4-17, but the diagram contains no alphabetic/numeric designations or test points. The components are identified only by their schematic symbols along with their electrical values or industry standard part designations. In the example shown, you know that the transistor is a 2N2222 type, and that the resistors have values of 470, 33k (33,000), 330k (330,000), and 680 ohms. The input capacitor has a value of 0.01 microfarad (µF), and the output capacitor has a value of 0.1 µF. The emitter capacitor, which goes across the 470-ohm resistor, has a value of 4.7 µF.

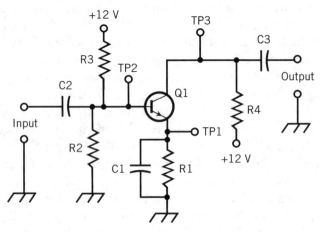

FIG. 4-17   *Schematic of an amplifier circuit that includes component designators and three test point (TP) locations.*

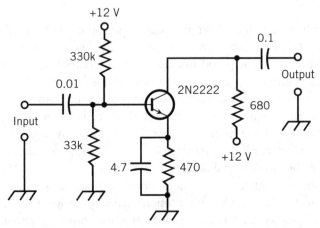

FIG. 4-18   *Schematic of the amplifier circuit of Fig. 4-17, but with component values rather than sequential designators. All capacitances are in microfarads (µF). All resistances are in ohms (Ω); k = 1000.*

Tip

In schematics like the one in Fig. 4-18, you'll often see a statement at the bottom that includes information about the units for the value designations. Such a statement might read "All capacitances are expressed in microfarads (µF). All resistances are expressed in ohms (Ω), where k = 1,000 and M = 1,000,000."

Schematic/Block Hybrids

Once in awhile you'll encounter a hybrid drawing that has blocks and schematic symbols combined. Figure 4-19 provides an example of this approach, which works well when you want to show the details of a particular circuit within a system and clarify that circuit's relationship to other parts of the system. The schematic portion of Fig. 4-19 shows a *buffer* of the sort that you'll sometimes find in a radio transmitter. The labeled blocks portray an oscillator (which precedes the buffer) and an amplifier (which follows the buffer).

This diagram serves two purposes. First, as you read the schematic portion, you can study the component makeup of the buffer circuit. Second, you get a good idea as to the buffer's place in the overall system relative to the other circuits. Figure 4-19 tells you that the buffer receives its input from a crystal-

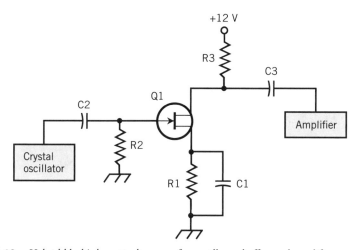

FIG. 4-19 *Hybrid block/schematic diagram of an oscillator, buffer, and amplifier system, showing the schematic details of the buffer circuit.*

controlled oscillator, and also that the buffer sends its output to an amplifier. Another schematic/block diagram combination might describe a different part of this same system.

Figure 4-20 is a hybrid diagram of the same system that Fig. 4-19 shows. In this version, you see the oscillator circuit in schematic detail, but the buffer and amplifier circuits are condensed into blocks. Figure 4-20 tells you, as Fig. 4-19 did, that the oscillator output goes to the buffer, and the signal then passes to the amplifier. As you look at Figs. 4-19 and 4-20 together, you can mentally envision all the system details up to the point where the signal enters the amplifier. The oscillator portion of Fig. 4-20 reveals the fact that the circuit can generate *Morse code* signals because it contains a *telegraph key*, which is a specialized switch (S1).

Now let's recall the block diagram of the power supply that you saw back in Fig. 2-1. Figure 4-21 comprises a schematic representation of that system (at A) along with a duplicate of Fig. 2-1 (at B). Drawing A portrays all the individual components in the power supply, while drawing B shows only the stages. Figure 4-22 highlights the schematic of Fig. 4-21A to reveal the contents of the blocks in Fig. 4-21B, but leaves out the alphabetic-numeric component designators to minimize clutter.

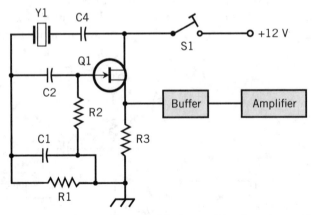

FIG. 4-20 *Hybrid block/schematic diagram of the system of Fig. 4-19, showing the schematic details of the oscillator circuit.*

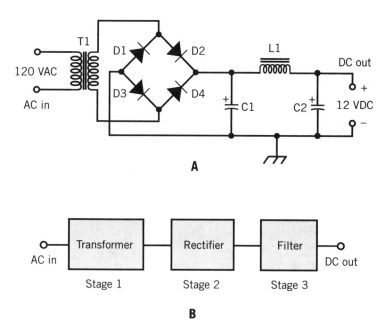

FIG. 4-21 *Schematic of a power supply (at A) such as the one in the block diagram you saw back in Fig. 2-1 (reproduced here at B).*

Follow the Flow

In the power supply, AC electricity enters at the left-hand end of Figs. 4-21A, 4-21B, and 4-22. The transformer steps the AC voltage down. The diodes in the diamond configuration constitute the full-wave bridge rectifier, which converts the AC to pulsating DC, taking advantage of both halves of the cycle. The ripple filter acts to "smooth out" the pulsations in the DC after rectification. The resulting *pure DC* proceeds to the output terminals at the right-hand end, where it appears as electricity of the sort that you'd find at the terminals of a 12-V battery.

A Vacuum-Tube RF Amplifier

The power supply diagrammed in Figs. 4-21 and 4-22 will operate the oscillator and buffer parts of the radiotelegraph transmitter of Figs. 4-19 and 4-20. But 12 VDC will not suffice for the amplifier that constitutes the final

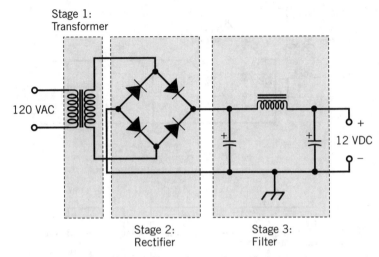

FIG. 4-22   *Here's how Fig. 4-21A relates to Fig. 4-21B. Component designators have been omitted to avoid clutter.*

part of the transmitter, because that amplifier employs a vacuum tube! You'll rarely see tubes nowadays except in amplifiers designed to provide high power output (hundreds or even thousands of watts). But "rarely" doesn't mean "never."

The circuit of Fig. 4-23 has a triode tube with an indirectly heated cathode. We omit the filament to keep the diagram from getting too messy. Within the circle that represents the tube envelope, the cathode symbol is at the bottom, the control grid symbol is in the middle, and the anode (or plate) symbol is at the top. Electrons flow generally from the cathode to the plate, passing through the control grid on the way. The control grid is a wire mesh or screen that interferes with the electrons to a greater or lesser extent, depending on the DC voltage that you place on it under no-signal conditions.

In most triode-tube amplifiers, you'll supply the control grid with a negative DC voltage relative to the cathode. You can get that voltage by giving R1, the control-grid resistor, a smaller ohmic value than R2, the cathode resistor, thereby elevating the cathode farther above electrical ground (to which the negative power-supply voltage goes) than the control grid. As you increase the negative *grid bias* by making R1 progressively smaller with respect to R2, the control grid impedes the flow of electrons more and more.

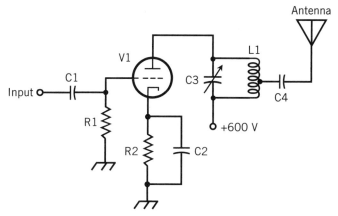

FIG. 4-23 *Schematic of the amplifier circuit in the radiotelegraph transmitter of Figs. 4-19 and 4-20. This "beast" employs a vacuum tube and needs a dedicated high-voltage (+600 V) DC power supply.*

Siblings?

You can imagine a triode tube as a "big sister" of an N-channel JFET. The tube's cathode corresponds to the source of the JFET; the tube's control grid corresponds to the gate of the JFET; the tube's plate corresponds to the drain of the JFET. Major differences between a triode tube and an N-channel JFET involve

- The physical construction of the electrodes
- The nature of the charge carriers
- The medium through which the charge carriers pass
- The power-supply voltage
- The ease with which you can find an appropriate tube

In a vacuum-tube RF power amplifier, you can apply the input signal to the control grid or to the cathode. In the circuit of Fig. 4-23, the input signal goes to the control grid. That arrangement allows the amplifier to work with far less input power than it would need if the control grid were grounded and the input signal were applied to the cathode. However, when you ground the cathode for RF (that's the function of C2) and apply the input signal to the control grid, you run the risk of unintended oscillation between the control grid and the plate. Then you get *spurious emissions*. That means trouble! In a so-called *grounded-cathode* circuit, you sacrifice *stability* in order to get greater *gain* and *sensitivity* than you could get with a *grounded-grid* circuit.

Follow the Flow

In the circuit of Fig. 4-23, the RF signal from the buffer enters through capacitor C1, which keeps DC in the amplifier from affecting the buffer while allowing the RF signal to pass easily. That signal goes to the control grid, whose DC bias is determined by resistors R1 and R2. Capacitor C2 allows some DC voltage to exist on the cathode but keeps it at RF ground. Small changes in the control-grid RF signal voltage cause large fluctuations in the electron flow (current) from the cathode to the plate. You adjust the variable capacitor C3 so that it, along with inductor L1, creates a condition of *resonance* in the output circuit. Engineers some-times call such a parallel inductance/capacitance (LC) arrangement a *tank circuit*. The RF signal voltage between the top and bottom of the tank circuit reaches a maximum at resonance. You adjust the position of the tap on L1 to match the antenna's *impedance* to the amplifier's output impedance, ensuring that the greatest possible amount of RF energy gets to the antenna. The RF output signal passes through capacitor C4 to the antenna and then into the universe beyond: maybe to a listener next door, in another state, on another continent, or (for all you know) on an unknown planet!

Three Basic Logic Circuits

Look back at the section on logic gates at the end of Chapter 3. Reexamine the schematic symbols (Fig. 3-58) and the descriptions of how the logic gates process the inputs to derive the output (Table 3-1).

You can break Table 3-1 down into logic functions where 0 equals "falsity" (the low state) and 1 indicates "truth" (the high state). Engineers and logi-cians call such arrays *truth tables*. Tables 4-2 through 4-7 restate the verbal descriptions in Table 3-1 as arrays of numeric logic states.

Let's create three schematic diagrams that show combinations of logic gates, and compile truth tables to show the logic states that those circuits produce.

TABLE 4-2   *Truth table for logical negation (NOT).*

X	NOT X
0	1
1	0

TABLE 4-3   *Truth table for the logical OR operation (inclusive).*

X	Y	X OR Y
0	0	0
0	1	1
1	0	1
1	1	1

TABLE 4-4   *Truth table for the logical AND operation.*

X	Y	X AND Y
0	0	0
0	1	0
1	0	0
1	1	1

TABLE 4-5   *Truth table for the logical NOR operation.*

X	Y	X NOR Y
0	0	1
0	1	0
1	0	0
1	1	0

TABLE 4-6   *Truth table for the logical NAND operation.*

X	Y	X AND Y
0	0	1
0	1	1
1	0	1
1	1	0

TABLE 4-7   *Truth table for the logical XOR operation (exclusive OR).*

X	Y	X XOR Y
0	0	0
0	1	1
1	0	1
1	1	0

Case 1

Suppose that you place logical inverters (NOT gates) in series with both inputs of an AND gate. Figure 4-24 illustrates this arrangement as a schematic. The inputs appear at points X and Y. Let's specify and label two intermediate circuit points as P and Q, and then call the output point R. At point P, you have the condition "NOT X." At point Q, you have the condition "NOT Y." At point R, you have the condition

$$(NOT\ X)\ AND\ (NOT\ Y)$$

When you compile all the possible logic states at each of these points, you'll get Table 4-8. This table shows the circuit condition at every point indicated for all possible input combinations:

- $(X,Y) = (0,0)$
- $(X,Y) = (0,1)$
- $(X,Y) = (1,0)$
- $(X,Y) = (1,1)$

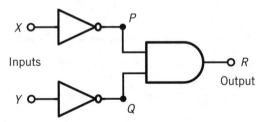

FIG. 4-24   *Logical inverters in series with both inputs of an AND gate.*

TABLE 4-8 *Truth table for the logic circuit of Fig. 4-24.*

X	Y	P	Q	R
0	0	1	1	1
0	1	1	0	0
1	0	0	1	0
1	1	0	0	0

Case 2

Suppose that you follow an XOR gate with an inverter. Figure 4-25 shows this sequence of logic gates in schematic form. As before, let's call the input points X and Y. At point P, you have the condition "X NOR Y," which is high when points X and Y have opposite logic states, and low when points X and Y have the same logic state. At point Q, you have the output state, which is always the opposite of the state at point P. Table 4-9 depicts the logic states at points X, Y, P, and Q for all possible input combinations:

- $(X,Y) = (0,0)$
- $(X,Y) = (0,1)$
- $(X,Y) = (1,0)$
- $(X,Y) = (1,1)$

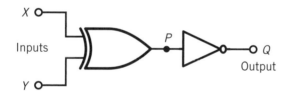

FIG. 4-25 *An XOR gate followed by a logical inverter.*

TABLE 4-9 *Truth table for the logic circuit of Fig. 4-25.*

X	Y	P	Q
0	0	0	1
0	1	1	0
1	0	1	0
1	1	0	1

Case 3

Suppose that you place an inverter in series with each input of an XOR gate. Figure 4-26 is a schematic of this arrangement. At point *P*, you have the condition "NOT *X*." At point *Q*, you have the condition "NOT *Y*." At point *R*, you have the condition

$$(NOT\ X)\ XOR\ (NOT\ Y)$$

Table 4-10 breaks down all possible logic states in this circuit. Once again, four possible input combinations exist, and you must scrutinize what happens as the signals make their way through the circuit for each input combination:

- $(X,Y) = (0,0)$
- $(X,Y) = (0,1)$
- $(X,Y) = (1,0)$
- $(X,Y) = (1,1)$

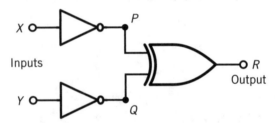

FIG. 4-26   *Logical inverters in series with both inputs of an XOR gate.*

TABLE 4-10   *Truth table for the logic circuit of Fig. 4-26.*

X	Y	P	Q	R
0	0	1	1	0
0	1	1	0	1
1	0	0	1	1
1	1	0	0	0

Rigor and Reassurance

You might protest against the process of breaking down a logic circuit, state by state and point by point, as needlessly tedious. But if you want to ensure that you get an accurate result, you should drag your mind through every step, one at a time. Make sure that you have a clear head and a patient mood, because this process (I admit) can get dull indeed, especially when you have three or more input variables to contend with!

Summary

This chapter has given give you some diagram-reading expertise, so you can read diagrams that show how circuits get assembled, or draw diagrams on the basis of known circuit details. After some more practice, you should have no trouble viewing a simple schematic and visualizing the finished circuit. You might even get an idea of how to arrange the components on a circuit board, even though schematics don't explicitly tell you how to do that.

Given time, schematic symbology will "grow on you"; it'll evolve into a new language in which you'll eventually learn to think, just as you can think in terms of words, mathematical equations, musical scores, architectural blueprints, Morse code, or any other language if you use those tools regularly and often.

Onward!

When you see a device or system represented as a schematic, you can visualize the component interconnections and perhaps make a few mental notes on the physical aspects of the circuit construction. But if you get the opportunity, you should open up the thing, look inside it, and operate it in real time.

5

Complex Circuits

As you learn to read and draw schematic diagrams, don't get discouraged by occasional difficulties. You might think, "Anyone can learn to read schematics of simple circuits, like those that have a transistor or two, a few capacitors, and a few resistors. But it'll take years before I can decipher complex schematics." Not so! You must put some effort into the learning process, but you can always break a complicated system down into simple circuits.

Identifying the Building Blocks

Even a system whose diagram looks overwhelming at first glance comprises smaller circuits interconnected in a straightforward way. A system with six diodes, ten inductors, ten transistors, and dozens of resistors and capacitors might resolve into four or five simple circuits, each containing only a few components. If you look at the entire system schematic all at once, you might as well try to eat a jumbo hamburger in a single swallow. With the diagram, as with the burger, you face an easier task if you assimilate the thing in little bites or pieces.

Figure 5-1 shows a "crystal radio" receiver built with an antenna, a tapped air-core inductor, a variable capacitor, an RF diode, and a fixed capacitor. The term "crystal" derives from the original construction of RF diodes in the early 1900s. In order to get a one-way current gate to act as a signal *detector* (or *demodulator*), radio experimenters placed a strand of fine wire called a *cat's whisker* into contact with a chunk of crystalline lead sulfide called *galena*. Today, semiconductor diodes process RF signals in the same way as "crystals" once did, even though modern RF diodes don't look like the old galena-and-cat's-whisker contrivances.

You can't call the circuit of Fig. 5-1 "complicated," but it performs some sophisticated tricks. Aside from the antenna (an external component such as a length of wire running outdoors under a window sash to a tree) and a sensitive

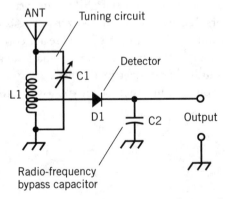

FIG. 5-1   *Schematic of the tuned circuit and detector stages of a "crystal radio" receiver. This circuit produces a weak audio-frequency (AF) signal.*

earphone or headphone that you can connect at the output terminals to hear radio stations (however faintly), this circuit contains only four components: the coil, the diode, and two capacitors.

You can connect an amplifier to the output of the "crystal radio" to boost the *audio-frequency* (AF) volume to the point where it can drive a headset to a comfortable listening level. Figure 5-1 doesn't include a headset at the output. To accomplish that feat, you'll need another circuit, along with a battery or other source of DC power to make the AF output signal strong enough.

Follow the Flow

In the "crystal radio" (Fig. 5-1), RF current from the antenna *resonates* (reverberates) in L1 and C1. The *resonant frequency* depends on the size of the coil and also on the setting of C1. That frequency determines which station (if any) that you hear. Diode D1 converts the RF current to pulsating DC that contains a tiny bit of both the RF and the AF energy from the original transmitter. Capacitor C2 shorts the RF part of the signal to ground, leaving only AF energy at the output. You must adjust the position of the coil tap, by trial and error, to maximize the AF output.

Figure 5-2 shows another fairly simple schematic: an AF amplifier that employs a single NPN bipolar transistor. In addition to the transistor, this circuit has four resistors and three capacitors for a total of eight components. It needs a source of DC power such as a battery (not shown), which provides 12 V. This circuit accepts a low-level AF signal (the output of a "crystal radio," for example) at the input terminals and boosts the power to a level strong enough to make audible sound come out of a headset. An experienced engineer might need a couple minutes to scribble the schematic and a couple of hours to build and test the circuit, "tweaking" the component values to get the best possible performance.

Because the circuit of Fig. 5-2 takes a weak signal and boosts it to a reasonable (but not very powerful) level, it's sometimes called a *preamplifier*. If you want to drive a speaker so that all the students in a classroom can hear the sound, you'll need more amplification. You can get that extra AF boost with one or more additional amplifiers connected to the output of the preamplifier.

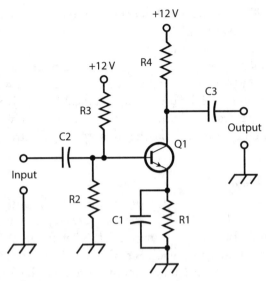

FIG. 5-2 *An AF preamplifier circuit that can work with the "crystal radio" to produce a signal strong enough to drive a headset.*

Follow the Flow

In the circuit of Fig. 5-2, AF current passes from the input through C2 to the base of Q1. Capacitor C2 keeps the power supply DC from affecting the behavior of the previous circuit (the "crystal radio"). Tiny current fluctuations at the base of Q1 cause larger current variations through the transistor from the emitter to the collector. The amplified AF signal passes through C3 to the output. The resistors govern the current flowing through Q1; you must choose their values by experiment for optimum amplification. Capacitor C1 keeps the emitter at AF signal ground while resistor R1 allows some DC voltage to exist there.

Figure 5-3 is a schematic of a circuit that looks, at first glance, more complicated than either Fig. 5-1 or Fig. 5-2. But as you examine Fig. 5-3 for a minute, you'll see that it's nothing more than the composite of the "crystal radio" (Fig. 5-1) and the AF preamplifier (Fig. 5-2). The components are re-numbered generally going from left to right, the direction of signal flow through the system. (You should never duplicate a component designator within a single schematic.) In Fig. 5-3, the connection between the original

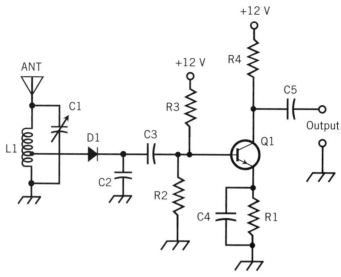

FIG. 5-3 *Combination of "crystal radio" tuned circuit, detector, and AF preamplifier stages. Some component designators are updated from Fig. 5-2.*

"crystal radio" and the preamplifier corresponds to the short, horizontal line that goes from the dot above C2 to the left-hand side of C3.

Now that you can envision the two building blocks that make up the circuit of Fig. 5-3, the whole diagram looks elementary, wouldn't you say? You can follow the signal as it passes through the "crystal radio" and then through the AF preamplifier. The entire process, from the RF signal arriving at the antenna to the AF energy appearing at the output, takes place in a tiny fraction of a second.

The circuit of Fig. 5-3 produces a stronger signal than the feeble output of the "crystal radio" alone, which gets its power only from RF current that flows in the antenna. Nevertheless, even the amplified AF output from the circuit of Fig. 5-3 isn't strong enough to provide a comfortable listening volume in a loudspeaker. It offers some sound power if you connect a headset to it, but not much. In order to further boost the AF signal level, you'll need a substantial *power amplifier.*

Figure 5-4 shows a two-transistor ensemble that can produce respectable sound power. It's called a *push-pull amplifier.* The top transistor amplifies half the AF waveform and the bottom transistor amplifies the other half. Imagine that Q1 "pushes" and Q2 "pulls" so when you combine their outputs, you get a magnified version of the entire input waveform. The push-pull amplifier can take the weak AF signal from a low-level amplifier (such as the circuit of Fig. 5-3) and boost it enough to make some loud sound come out of a speaker!

Cats and Tigers

Whenever you hear the term *power amplifier*, remember that the circuit does exactly what its name implies: It accepts a signal that can yield a certain amount of *power* and turns it into a signal that can yield more *power*. If a circuit amplifies current, it doesn't necessarily amplify power. The same holds true for voltage; voltage amplification does not always translate into power amplification. If you call current or voltage amplifiers "mewing cats," you might describe power amplifiers as "roaring tigers"!

If the input signal doesn't contain much power, then the circuit of Fig. 5-4 won't get enough *drive* (input power) to produce a decent output signal. The circuit of Fig. 5-3 (the AF preamplifier) can provide enough "oomph" to adequately drive a push-pull AF power amplifier such as the one diagrammed in Fig. 5-4. The circuit of Fig. 5-1 (the "crystal radio" alone) can't.

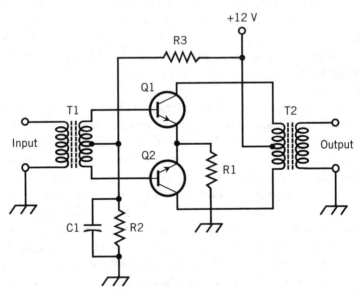

FIG. 5-4   *An AF power amplifier circuit suitable for driving a speaker.*

Follow the Flow

In the circuit of Fig. 5-4, the input signal goes through an AF transformer, T1, and appears across its secondary. Transistor Q1 handles half of the cycle and Q2 handles the other half, as the AF input current flows up and down through the secondary of T1. Transformer T2 assembles the two halves of the amplified cycle back into a complete AF signal stronger than the one entering T1. When you connect a speaker to the secondary of T2, you'll hear some loud sound!

If you combine the circuits of Figs. 5-3 and 5-4 in *cascade* (one after the other), you get a complete AM radio receiver that will produce respectable sound from a speaker. Figure 5-5 shows the entire three-transistor AM radio receiver in a single schematic. Again, we've had to change some of the component designators that appeared in previous diagrams, so they increase generally as you go from the original input at the antenna to the final output at the speaker, taking pains to ensure that you don't inadvertently give two different components the same name.

So There!

If you had seen Fig. 5-5 at the start of this chapter, you might have gotten confused and frustrated. But now that you can envision how the building blocks go together, you know that you don't have to "choke down the whole burger in one bite."

The evolution of an electronic system breaks down into a well-defined sequence. First, the individual components (resistors, capacitors, diodes, and so on) combine to form simple circuits. Second, those simple circuits combine to make complex devices or, in some cases, the whole system. Third, if the design is sophisticated, the complex circuits combine to form the complete system.

Several different devices (some of them simple and others not so simple) can combine to form a large system. An amateur radio station offers a good example. It might have a *transceiver* (transmitter/receiver in a single box), an *antenna tuner*, a *personal computer*, an *interface* unit that goes between the computer and the transceiver, a microphone that lets you transmit voice signals, a *speech processor* that goes between the microphone and the transceiver, a

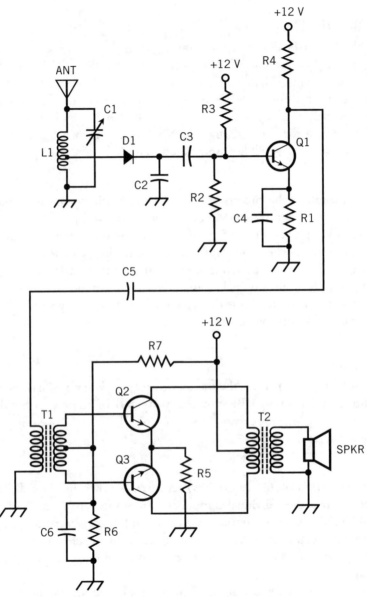

FIG. 5-5   *Complete radio receiver circuit. Some of the component designators in the AF power amplifier stage are updated from Fig. 5-4. This schematic includes the speaker.*

key that lets you transmit in Morse code if you want, a headset, a speaker, and a power supply that converts utility AC into the DC from which the whole system gets the various forms of electricity that it needs in order to function.

Page Breaks

Figure 5-5 is a "respectably complicated" diagram. You can draw the system in a two-level format with the detector and preamplifier on top, and the audio power amplifier on the bottom. A long, tortuous line, broken in the middle by C5, represents the connection between the preamplifier output and the power amplifier input. There's nothing technically wrong with this diagram, but some people would rather see it all on one level. In order to render the diagram that way, you could make it extremely small, or else draw it sideways on the page. You could even produce it on a foldout page (the sort of thing that they do in those upscale print magazines when they want to show you something spectacular).

You have yet another alternative, though! You can split the diagram into multiple pages. You need not use that approach with the schematic of Fig. 5-5, but when you get to extremely complicated systems such as amateur radio transceivers, television sets, or computers, you might want to take advantage of that option. Figure 5-6 shows how you can use this technique with the diagram of the radio receiver from Fig. 5-5. Figure 5-6A puts the tuner, detector, and AF preamplifier right-side-up on a single page along with an output designator that appears as an X inside an arrow that points off the page toward the right. Figure 5-6B shows the push-pull AF power amplifier with an input designator comprising an X inside an arrow that points off the page toward the left.

Tip

In Figs. 5-6A and 5-6B, the wedge-like arrows represent points that you must connect directly to each other when you build the system. These arrows take the place of the long, tortuous line in Fig. 5-5. In this case you need only one pair of arrows. Some circuits need two or more pairs; you might label them as X, Y, and Z, for example. When you build the system, you'll connect the pairs of points represented by the two X's, the two Y's, and the two Z's.

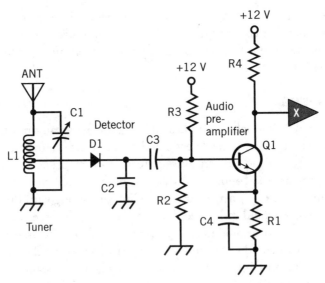

FIG. 5-6A   *Tuner, detector, and AF preamplifier stages in the radio receiver. The wedge X represents an extension to drawing B.*

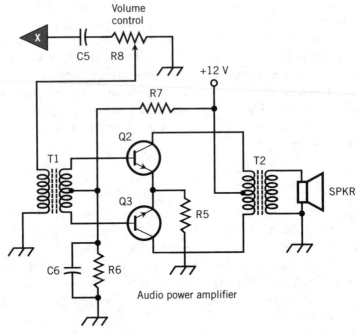

FIG. 5-6B   *The AF power amplifier and speaker in the radio receiver. The wedge X represents an extension from drawing A.*

Let's follow the signal through Fig. 5-6A. A radio wave in free space causes RF current to flow in the antenna, and also through inductor L1. Capacitor C1 causes the inductor/capacitor combination (called an *LC circuit*, where an italic *L* stands for "inductance" and an italic *C* stands for "capacitance") to resonate at the frequency of the RF signal that you want to hear. Diode D1 detects (demodulates) the signal, splitting the AF and RF portions apart. Capacitor C3 passes the AF portion of that signal along to the base of transistor Q1. Capacitor C2 shunts (short-circuits) the RF portion of the diode's output to ground, because the circuit doesn't need the RF energy anymore; its presence could, in fact, cause trouble in the following stages! Transistor Q1 acts as an amplifier for the weak AF signal at its base. Resistors R1, R2, R3, and R4 ensure that Q1 gets optimum DC voltage (called *bias*), so that it will produce the greatest possible AF gain. Capacitor C4 keeps the emitter at AF signal ground while allowing some DC voltage to exist there. The AF output signal, along with some DC from the power supply ($+12$ V), goes off the page through the rightward-pointing arrow marked X.

Now let's look at Fig. 5-6B and follow the signal after it comes in from Fig. 5-6A. The AF energy, along with some DC, appears at the leftward-pointing arrow marked X. Capacitor C5 blocks the DC so that only the AF current can reach potentiometer R8.

Whoa!

Have you noticed that R8 did not exist in previous diagrams of this receiver? It serves as a volume control.

The full AF voltage appears across the entire resistance of R8. The arrow touching the zig-zag symbolizes the potentiometer *wiper* or *slider*, which "picks off" AF voltages that can range from zero (all the way to the right-hand end of the zig-zag, at ground) to the maximum possible (all the way to the left-hand end of the zig-zag, at C5). The slider signal causes AF current to flow in the primary of transformer T1. From there, the signal flows as described in "Follow the flow" for Fig. 5-4. The only difference between this situation and that one is the numbering of the component designators. You can also add a speaker to the output of T2, as you did in Fig. 5-5.

Some More Circuits

Figure 5-7 shows an antenna matching circuit known as an *L network*. In this case, the letter L refers to the general layout of the components in the diagram, not to the inductor in the circuit. (Actually, to make the coil and capacitor in Fig. 5-7 take the shape of an uppercase L in the layout, you'll have to rotate the page 90 degrees clockwise and then hold it up to a mirror! But you get the general idea, right?)

Figure 5-8 shows another type of antenna matching network, which comprises the circuit from Fig. 5-7 with an extra capacitor added at the input end. Engineers call this type of circuit a *pi network* because its components, in the schematic layout, resemble the shape of the upper case Greek letter pi (π).

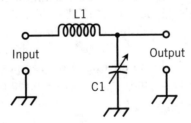

FIG. 5-7   *An L network comprising an inductor and a variable capacitor.*

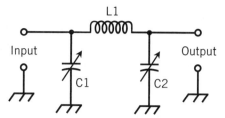

FIG. 5-8 *A pi network comprising an inductor and two variable capacitors.*

Figure 5-9 shows a circuit that's more complicated than the ones in Figs. 5-7 and 5-8, but in a sense contains them both put together. When you follow a pi network with an L network, you get a *pi-L network*. The advantage of a circuit like the one in Fig. 5-9, compared to those in the previous two diagrams, lies in its ability to make RF transmitters work with antennas that otherwise wouldn't accept power. Those two extra components can go a long way!

Tip

You can make the circuits of Figs. 5-7, 5-8, and 5-9 more versatile by using adjustable coils such as roller inductors, which have grown popular among radio amateurs in recent years. You learned a little bit about them in Chapter 3. Just for fun, do an Internet search on "roller inductor" by entering the term in the phrase box of your favorite search engine. You should find some photographs of these components. They allow for precise adjustment of inductance, and some of them have calibrated crankshafts so that you can easily reset them to any previous position.

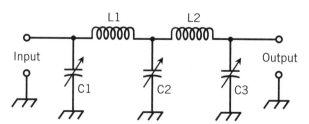

FIG. 5-9 *A pi-L network comprising two inductors and three variable capacitors.*

If you're old enough, you'll remember the days when you had to learn the *International Morse code* (often simply called "the code") to get an amateur radio operator's license. That mandate has passed into history, but some amateur radio operators still enjoy communicating this way. If you want to do that, you must learn to "read" and "speak" in the code. To that end, you can build a *code practice oscillator* such as the one diagrammed in Fig. 5-10. It's an AF oscillator that you can switch on and off with a *straight key* or *telegraph key*, labeled "Key" in the figure. (Because a telegraph key technically constitutes an SPST switch, you could label it S1.)

When you first examine Fig. 5-10, you might wonder why an AF oscillator circuit needs so many components. Can't you build a simple AF amplifier like the preamplifier or power amplifier discussed earlier, and feed some of

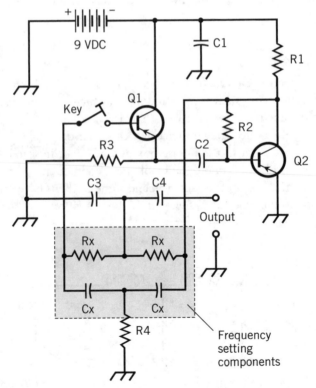

FIG. 5-10 *An AF code-practice oscillator using two PNP bipolar transistors. The values of Rx and Cx determine the frequency.*

the output back to the input? Well, yes, you can do that; but if you want a decent "tone" to come out of your code practice oscillator, you'll get superior results with a circuit like the one in Fig. 5-10. It's called a *twin-T oscillator* because of the T-shaped configurations including the resistors marked Rx and the capacitors marked Cx. The twin-T oscillator produces a musical note that's pleasing to the ears and has a predictable and stable *pitch* (frequency).

Follow the Flow

The AF signal in the circuit of Fig. 5-10 goes around and around; that's how oscillation happens. You can start anywhere and follow the signal through the circuit, ending up back where you started. If you begin at the key, the signal goes into the base of Q1 where it gets amplified and undergoes *phase inversion* (the wave gets turned upside down). You take the output of Q1 from the emitter rather than from the collector to ensure a stable and reliable circuit. The Q1 output signal goes into the base of Q2 where it gets amplified and inverted again, so it comes out of Q2 in *phase coincidence* (reinforcing rather than opposing) with the signal at the key. The signal emerges from the collector of Q2 and goes down to the twin-T network comprising resistors Rx and capacitors Cx, whose values determine the oscillation frequency. From there, the signal goes back to the key and begins another round trip. You take the output from the upper part of the T network through capacitor C4.

The circuit of Fig. 5-10 uses a 9-V battery as its power source. In this example, the transistors are of the PNP type, so the collectors get a negative voltage while the positive battery terminal goes straight to ground. This circuit therefore constitutes a *positive-ground system*.

You might want to build a power supply from the AC utility mains rather than relying on a battery. If you want to do that, you'll need to design the supply so that it produces a negative voltage with respect to ground. Figure 5-11 shows a power supply that can provide pure, constant −9VDC. It's almost the same circuit as the one you saw back in Fig. 4-13, with the following exceptions:

- All the diodes go in the opposite direction (including the Zener diode).
- The electrolytic capacitor has the opposite polarity.
- The circuit produces a lower DC output voltage.

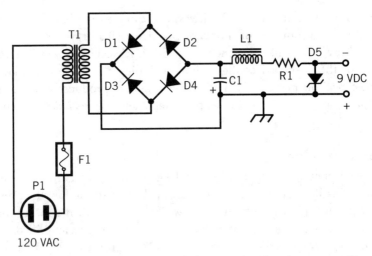

FIG. 5-11 *A regulated −9 VDC power supply that can power the code-practice oscillator of Fig. 5-10. Note the positive ground for use with PNP transistors.*

Figure 5-12 shows a complete AF code practice oscillator system that can operate from the AC utility mains. It combines the power supply of Fig. 5-11 with the oscillator of Fig. 5-10. As with the radio receiver schematic of Fig. 5-5, you connect the power supply output to the oscillator in Fig. 5-12 with a single line (but not as long as the one in Fig. 5-5). Figure 5-12 also shows a volume control (R6) and a pair of headphones.

You might want to boost the twin-T oscillator output so that the AF power will drive a loudspeaker, letting you send Morse code to a classroom full of students! A push-pull AF amplifier, such as the one you used for the radio receiver except with PNP rather than NPN transistors, will do this job. Then you'll get a system with three essential circuits: a power supply, an oscillator, and an amplifier. Figures 5-13A, B, and C show the complete system schematic spread across three different pages.

Heads Up!

As you examine Figs. 5-13A, B, and C, note the designators inside the arrows. The right-pointing arrow X in Fig. 5-13A goes to two places on subsequent pages: the left-pointing arrow X in Fig. 5-13B and the left-pointing arrow X in Fig. 5-13C. The arrows marked Y in Figs. 5-13B and C each represent one connection.

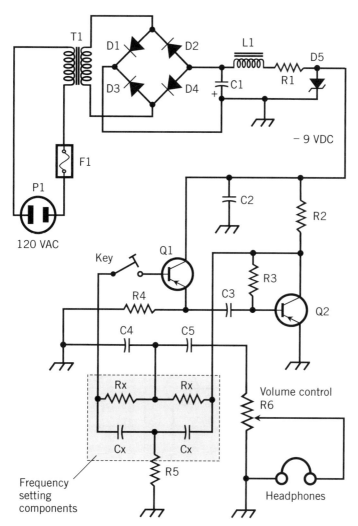

FIG. 5-12 *Combination of the regulated power supply and code practice oscillator. Note the addition of the volume control and headphones.*

Figure 5-14 shows a simple LC circuit that resembles the L network of Fig. 5-7, but the inductor and capacitor have changed places, the capacitor is fixed rather than variable, and the inductor has a powdered-iron core instead of an air core. In addition, this circuit performs a different function than the other one does. The circuit in Fig. 5-7 works mainly to tune an antenna system or to match the output of a transmitter to a particular antenna. The circuit in Fig. 5-14 is designed to let signals get through (or not) depending on their

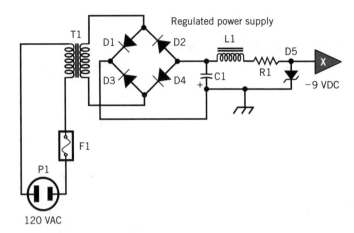

FIG. 5-13A *A regulated power supply for a classroom code-practice system. The wedge X represents an extension to illustrations B and C.*

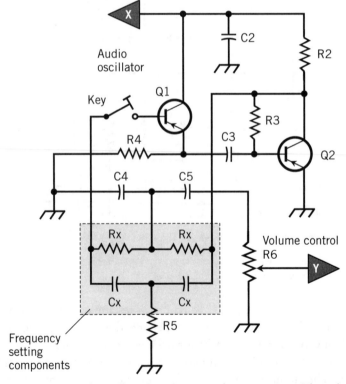

FIG. 5-13B *A twin-T audio oscillator for a classroom code-practice system. The wedge X represents an extension from illustration A on the previous page. The wedge Y represents an extension to illustration C.*

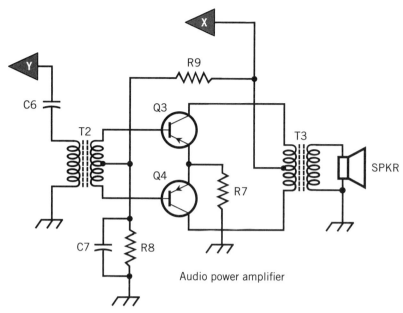

FIG. 5-13C *An AF power amplifier for a classroom code-practice system. Note the PNP transistors, consistent with the negative power-supply voltage (positive-ground system). The wedge X represents an extension from illustration A on the previous page. The wedge Y represents an extension from illustration B.*

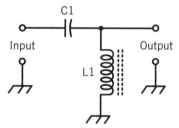

FIG. 5-14 *A simple frequency-sensitive filter circuit.*

frequency. It's called a *highpass filter* because it lets signals pass more easily as the frequency increases. The exact frequency at which high attenuation (lots of signal loss) changes to low attenuation (little or no signal loss) depends on the values of the capacitor and inductor.

Figure 5-15 shows a more complicated LC circuit that comprises two filters in cascade. The first filter, made up of the series-connected capacitor C1 and the parallel-connected inductor L1, has the same design as the one in Fig. 5-14. The second filter, made up of the series-connected inductor L2 and the parallel-connected capacitor C2, forms a *lowpass filter*. It works in the opposite manner from a highpass filter, letting signals get through more easily as the frequency goes down.

When you follow a highpass filter with a lowpass filter, and if you choose the *cutoff frequencies* (or transition points) so that the highpass filter cutoff frequency lies below the lowpass filter cutoff frequency, you get a *bandpass filter*, which a signal can pass through easily when its frequency lies between the two cutoffs. If the frequency lies outside the "zone of easy passage," it gets greatly attenuated.

Tip
You'll often find circuit configurations repeated in electronic system design. Sometimes you'll see several identical or similar circuits connected together. Sometimes their components all have the same values, and sometimes they don't. Sometimes the circuits connect in series (end-to-end, like the links in a chain); sometimes they connect in parallel (across each other, like the rungs in a ladder).

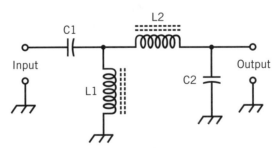

FIG. 5-15   *A complex frequency-sensitive filter comprising two simple, but different, filters connected in cascade.*

If you know how one of the circuits in a repetitive system operates, then you know, by extension, how all the circuits work. A problem that develops in one circuit might also arise in any of the others. For example, suppose that you learn (by testing) that an oscillator has changed frequency because of a defective resistor in the base circuit. If another oscillator of the same configuration develops the same malfunction, you can consult the schematic, locate the second base resistor, and conduct some tests to see if it, too, has gone bad. Without the schematic, you'd have a difficult time locating the rogue resistor.

Bang for the Buck

A single component that costs a few pennies can, all by itself, bring down a sophisticated system that costs hundreds of dollars. That's why electronics technicians deserve good pay for their work! (Alas, they don't always get it.)

If you break a gigantic, malfunctioning system down into multiple complex circuits, you can figure out which circuit might be responsible for the problem. You can then split the complex circuit into simple ones and figure out which circuit is most likely the culprit. Within that rogue circuit, you can examine the components one at a time. If you follow such a process of elimination with the help of schematics, you can repair the equipment with less trouble than you'd go through if you didn't have the diagrams. Whenever you go through this process, an engineer would say that you *troubleshoot to the component level*.

You're Hired!

If you can troubleshoot complicated electronic systems to the component level, you'll find yourself in high demand. All you'll have to do is show up on time when someone calls you for service, do the job right, leave, and send your grateful customer a bill. When word gets around about your competence and reliability, you'll have a guaranteed job for the rest of your life.

Many electronic system failures arise from a problem with a single component. Sometimes this glitch will cause other components to go bad, too. Once in a while you'll discover two or more defective components causing a single problem. You must familiarize yourself with the system, and then, after you've studied and understood the system's normal operating state, you can use a schematic to get a good idea of where future troubles might arise. By following this procedure whenever malfunctions occur, you can identify the most suspect individual component(s). Then you need only find those component(s) in the physical system, get into the equipment with the appropriate test instrument, and check the suspect component(s) one by one.

Tip
Even if you have good test equipment, you'll find it difficult to quickly identify defective sections (and especially individual components) without a schematic, because you might not know where on the circuit board or chassis to look!

Getting Comfortable with Large Schematics

No matter how much you enjoy electronics, you can't expect to sit down as a beginner and read complicated schematics with ease. You need to climb a learning curve. First of all, you must make certain that you know every schematic symbol that you expect to see. Complex schematics can serve as a great learning tool, because they contain lots of symbols, some of which you probably won't know at first. You can use these diagrams to help you learn the symbols.

Once you feel comfortable with the individual symbols, put away the complex schematics and start looking over diagrams of simple, common circuits. You'll find lots of them in magazines for electronics enthusiasts. (You'll also find plenty of simple projects and related diagrams in *Electricity Experiments You Can Do at Home*, published by McGraw-Hill.) Don't limit your studies to one type of schematic, such as those that portray only amplifiers. Check into oscillators, power supplies, solid-state switches, RF circuits, AF circuits, and anything else you can find. You'll discover similarities among different types of circuits, sometimes with no significant differences other than a few changes in component values. When you can identify an amplifier or oscillator or detector by looking at its schematic, then you'll know that you've made progress.

Don't Hurry!
Once you can identify simple electronic circuits by looking at their schematics, you should move on to more complicated drawings. But if you try to advance too quickly, you might grow frustrated and give up altogether. Take plenty of time!

Your next step will include devices that combine a few of the simple circuits you've previously studied. Sometimes you'll see additional components that electrically match the output of one circuit to the input of another. Choose books and publications that offer both theoretical and practical discussions of the circuit that the schematic depicts. Even better, build some simple circuits in a home workshop.

You might get a surprise when you see how your first "homebrew" electronic circuit looks in real life when compared with the schematic. Your study will continue from this point by examining the functional circuit and noting the relationship of the physical components to those in the schematic. You can expand your electronics knowledge by experimenting with these circuits (substituting different components, for example). You might find a way to make the circuit work better than it originally did. Note the improvements that you make, and draw a new schematic that reflects them all. If you didn't make many changes, you can pencil in the changes on the schematic from which you built the original circuit.

When you feel comfortable building simple circuits from schematics, you might want to combine two or more circuits to make a more sophisticated device. Take two schematics from a "projects" book and combine them on paper. You'll have to draw your own schematic to serve as a plan for the building procedure. You might know enough by this time to design and build a circuit that can *interface* the two (connect the output of the first circuit to the input of the second one so they both work at their best). When you combine circuit-building with the task of learning to read and create schematics, you'll improve your electronics knowledge more easily than you can do by merely looking at, and drawing, the diagrams.

Tip
You might grow bored as you pore over schematics for hours on end. But when you can refer to a portion of a schematic and then wire the components in place, much of your boredom will evaporate. If you're a budding *technophile* (technology lover), the stuff will get downright fascinating.

Before you know it, you'll have a solid knowledge of schematics and circuit-building. The circuits that you once imagined as complicated will seem elementary. Nevertheless, you should remain inquisitive. You might feel the temptation to stay with the types of circuits that you know best, and not venture into new territory. Don't let laziness get the better of you! As soon as you reach one stage of comfort, move on to more difficult diagrams. Keep building more complex projects. Of course, this practice can grow expensive if you overdo it, so if you can't build everything in sight, keep reading schematics and deciphering diverse circuit components anyway.

You'll forever stay amazed at what you know and what you don't know. For instance, many electronics neophytes imagine that a commercial AM radio transmitter must be a highly complex system. Most electronics novices are astonished to learn that the AM transmitter is less complicated than an old-fashioned transistorized pocket receiver that you might use to intercept the broadcasts. A commercial radio transmitter is a rather simple system, even if it's as big and massive as your car. The transmitter size directly correlates with the component size, which in turn directly relates to the amount of power that the system consumes. The power-supply transformer for a commercial transmitter, all by itself, might weigh as much as your home refrigerator! This and other components make the commercial broadcast transmitter large and heavy. The power-supply transformer for a small amateur radio transmitter will likely mass less than a kilogram. Nevertheless, both transformers will look the same in schematics.

Remember
System complexity bears little or no relation to physical size and weight. Complexity depends on the number of components as well as on the number of circuits that the system contains. A system larger than your house might be simple; a system smaller than your thumb might be complicated.

The most massive pieces of equipment are rarely the most complicated ones, both electronically and schematically. The tiny units that you can hold in the palm of your hand often take the prize for schematic complexity when you break them down to the component level. A tablet computer offers an excellent example. The integrated circuits (ICs or chips) inside such a system can contain millions of individual diodes, transistors, capacitors, and resistors. For this reason, an electronics novice should not shy away from any particular circuit, device, or system just because its size suggests complexity. You might be wrong, but even if you're right, every schematic will contain portions that you can comprehend.

After you've gotten past the intermediate stage of learning schematics, then you can tackle complex circuits, devices, and systems. You can break them down into multiple-circuit stages or devices, and ultimately into simple circuits. Try to obtain schematics of a complex nature that offer a complete and detailed explanation of how the circuits work.

Recall the block diagram of Fig. 2-2, the strobe light circuit that you saw in Chapter 2. Compare it to Fig. 5-16, a schematic that shows all the individual components. The whole diagram is tilted on its side, allowing it to fit on the page neatly. The circuit gets powered with 120 VAC, which enters at the left side of the schematic (after you rotate the page to make the diagram appear right-side up). The three terminals of the AC line take three separate paths along color-coded wires. A black wire goes to the fuse, a white wire goes to the power supply and timing components, and a green wire, coming from the "third prong" of the plug, goes to a substantial earth ground.

Follow the Flow

Starting with the plug, current passes through the fuse F1 when you close the switch S1. Current then goes through the rectifier diodes D1 and D2 in one direction only. From the diodes, one path goes all the way across the top wire to the "A" terminal of the strobe light, and the other path goes to the timing components comprising the resistors R1 through R4, the capacitors C1 and C2, and the neon lamp NE1. The timing components determine how long the lamp stays off between flashes. The silicon-controlled rectifier SCR1 performs the switching operation for the strobe lamp. The resistors R2 and R3, besides assisting with the timing, provide a junction where the bottom portion of the circuit interacts with the top portion, sending the proper signal through resistor R4 to the transformer T1 and finally to the strobe lamp terminals "T" and "K."

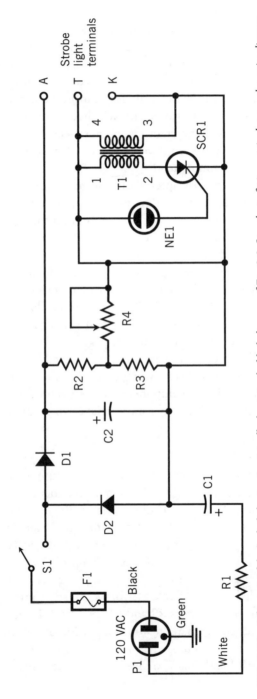

FIG. 5-16 *Schematic of the strobe light circuit originally shown in the block diagram of Fig. 2-2. In order to fit it on a single page, the entire diagram has been rotated counterclockwise by a quarter turn (90 degrees).*

126

Figures 5-17A and B show the same circuit as Fig. 5-16 does, except that the diagram is split into two sections so that it can all go right-side-up. The first part (Fig. 5-17A) shows the power supply and some of the timing circuitry. The second part (Fig. 5-17B) shows the frequency-adjusting potentiometer R4 along with the rest of the timing circuitry, the switching device, and the transformer that provides the strobe light with the voltage that it needs. The three right-pointing wedges in Fig. 5-17A connect directly to their left-pointing counterparts in Fig. 5-17B.

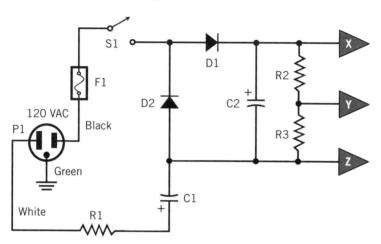

FIG. 5-17A *The plug, fuse, and rectifier portions of the strobe light circuit. Wedges X, Y, and Z represent extensions to illustration B.*

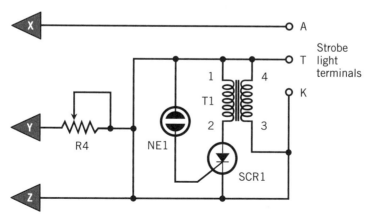

FIG. 5-17B *The timing and transformer portions of the strobe light circuit. Wedges X, Y, and Z represent extensions from illustration A.*

Op Amp Circuits

Now turn your attention back to Chapter 3 and re-read the section on op amps. After you're finished, let's take a look at a few real-world circuits that use these venerable little chips. All of these circuits are designed to work in the AF range.

Figure 5-18 shows an op amp wired up as a *non-inverting broadband amplifier*. The signal comes into the non-inverting input while negative feedback flows through the inverting input. You can add a resistor between the non-inverting input and ground, as shown in this schematic, to limit the input impedance and provide some extra stability to the amplifier. (You need not include this resistor if external components determine the input impedance, or if you want to keep the input impedance as high as possible.) In any non-inverting broadband amplifier, the output wave emerges in phase coincidence with the input wave over a wide range of frequencies. You connect a fixed resistor between the inverting input and ground, and another resistor, which can have either fixed or variable value, between the output and the inverting input.

Figure 5-19 shows an op amp serving as an *inverting broadband amplifier*. You'll find this arrangement in many of the same scenarios as you see non-inverting amplifiers. It resembles the circuit of Fig. 5-18, except that the input signal goes to the inverting input rather than the non-inverting input. You can add a resistor between the input terminal and ground to limit the input

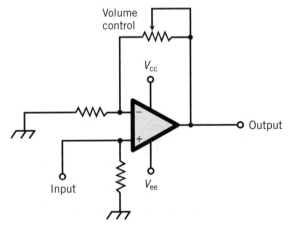

FIG. 5-18 *A variable-gain, broadband AF amplifier circuit that uses an op amp. This circuit produces its output signal in phase coincidence ("right-side up") with respect to the input signal.*

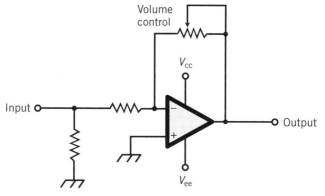

FIG. 5-19 *Another variable-gain, broadband AF amplifier circuit using an op amp. This circuit produces its output signal in phase opposition ("upside down") with respect to the input signal.*

impedance and enhance the stability, just as you can do with a non-inverting amplifier. (You won't need this resistor if external components determine the input impedance, or if you want to maximize the input impedance.) In an inverting amplifier, the output wave emerges in phase opposition with respect to the input wave. You can connect either a fixed resistor or a potentiometer between the output and the inverting input, exactly as you would do with the circuit of Fig. 5-18.

Follow the Flow

In the examples of Figs. 5-18 and 5-19, the potentiometer allows you to vary the amount of negative feedback, so it functions as a volume (AF gain) control. As you *increase* the potentiometer resistance, the gain *goes up* because the negative feedback *decreases*. As you *decrease* the potentiometer resistance, the gain *goes down* because the negative feedback *increases*.

An *inverting differentiator* is a circuit whose instantaneous output level varies in proportion to an *upside-down* version of the rate of change in the input signal level as a function of time. This arrangement produces an output signal with the same frequency as that of the input signal, although the waveform might (and often does) differ. Figure 5-20 shows a schematic of an op amp wired up as an inverting differentiator. This circuit provides some gain.

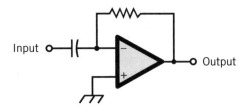

FIG. 5-20 *An op amp circuit that inverts and differentiates a signal and provides some gain.*

Thinning the Crowd

The V_{cc} and V_{ee} connection symbols don't appear with the op amp in Fig. 5-20. In schematics, engineers often omit these terminals to minimize clutter. Let's leave them out of our op-amp schematics from now on.

An *inverting integrator* is a circuit whose instantaneous output level varies in proportion to an upside-down version of the accumulated input signal level as a function of time. The output signal might have the same frequency as the input signal, but not necessarily. (In some cases, the output differs drastically from the input!) Figure 5-21 shows a schematic of an op amp wired up as an inverting integrator. As with the circuit of Fig. 5-20, this arrangement provides some gain.

Sinusoids

If you input a pure AC *sinusoid* (sine wave) to either of the above mentioned op amp circuits, you'll see another pure AC sinusoid at the output. In the circuit of Fig. 5-20, the output sinusoid gets amplified and shifted a quarter cycle to the *right* (later in time), so the output *lags* the input by 90 degrees. In the circuit of Fig. 5-21, the output sinusoid gets amplified and shifted a quarter cycle to the *left* (earlier in time), so the output *leads* the input by 90 degrees. (Remember, these circuits not only differentiate or integrate the sinusoids, but invert them also!)

If you connect specialized groups of resistors and capacitors together with op amps, you can create frequency-sensitive AF filters that can also provide amplification. Engineers call them *active filters* because they need a source

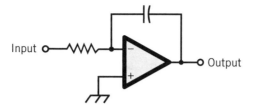

FIG. 5-21 *An op amp circuit that inverts and integrates a signal and provides some gain.*

of DC electricity (such as a battery) in order to work. All the way back in Chapter 3, you saw four gain-versus-frequency graphs showing:

- A *lowpass response* that favors low frequencies (Fig. 3-47A).
- A *highpass response* that favors high frequencies (Fig. 3-47B).
- A *resonant peak* that has maximum gain at a single frequency (Fig. 3-47C).
- A *resonant notch* that has minimum gain at a single frequency (Fig. 3-47D).

Figure 5-22 shows how you can wire up an op amp to produce a lowpass response. The values of resistor R and capacitor C determine the cutoff frequency (where the voltage gain is –3 dB, representing roughly 70.7 percent of maximum). The cutoff frequency drops if you:

- Increase the resistance but leave the capacitance constant.
- Increase the capacitance but leave the resistance constant.

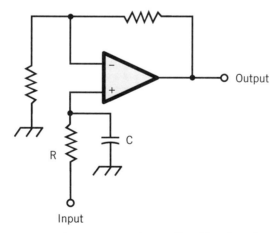

FIG. 5-22 *An op amp wired up to serve as a lowpass filter. The values of resistor R and capacitor C determine the cutoff frequency. This circuit provides some gain.*

- Increase both the resistance and the capacitance.

The cutoff frequency rises if you:

- Decrease the resistance but leave the capacitance constant.
- Decrease the capacitance but leave the resistance constant.
- Decrease both the resistance and the capacitance.

Figure 5-23 is a schematic of an op amp, a resistor R, and a capacitor C connected to produce a highpass response. As with the lowpass filter, the cutoff frequency drops if you:

- Increase the resistance but leave the capacitance constant.
- Increase the capacitance but leave the resistance constant.
- Increase both the resistance and the capacitance.

The cutoff frequency rises if you:

- Decrease the resistance but leave the capacitance constant.
- Decrease the capacitance but leave the resistance constant.
- Decrease both the resistance and the capacitance.

In order to calculate the cutoff frequency f (in hertz) of the lowpass filter of Fig. 5-22 or the highpass filter of Fig. 5-23, you need to know the resistance R in ohms and the capacitance C in farads. Then you can use the formula

$$f = 1 / (2\pi RC)$$

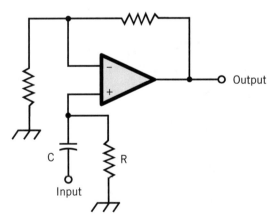

FIG. 5-23   *An op amp wired up to serve as a highpass filter. The values of resistor R and capacitor C determine the cutoff frequency. This circuit provides some gain.*

where π represents the ratio of a circle's circumference to its diameter (approximately 3.14159). This formula will also work if you express *R* in megohms and *C* in microfarads. (The frequency *f* will still come out in hertz.)

Italics or Not?

In conventional notation, you should use uppercase non-italic letters R and C to represent the words "resistor" and "capacitor," but uppercase *italic* letters *R* and *C* to represent the words "resistance" and "capacitance." You'll always want to use an *italic* lowercase letter *f* to represent the word "frequency."

Figure 5-24 is a schematic of an op amp, two resistors R1 and R2, and two capacitors C1 and C2 connected to produce a resonant peak response. The resonant peak (maximum gain) frequency drops if you:

- Increase the resistances but leave the capacitances constant.
- Increase the capacitances but leave the resistances constant.
- Increase both resistances and both capacitances.

The resonant peak frequency rises if you:

- Decrease the resistances but leave the capacitances constant.
- Decrease the capacitances but leave the resistances constant.
- Decrease both resistances and both capacitances.

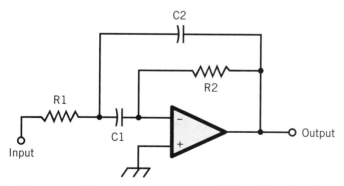

FIG. 5-24 *An op amp wired up to serve as a resonant peak filter. The values of resistors and capacitors R1, R2, C1, and C2 determine the resonant frequency.*

To calculate the resonant peak frequency f (in hertz), you need to know the resistance values $R1$ and $R2$ in ohms and the capacitance values $C1$ and $C2$ in farads. Then you can use the formula

$$f = 1 / [\, 2\pi\, (R1\ R2\ C1\ C2)^{1/2}\,]$$

This formula will also yield f in hertz if you express both resistances in megohms and both capacitances in microfarads.

Figure 5-25 is a schematic of an op amp, four resistors R1 through R4, and two capacitors C1 and C2 connected to produce a resonant notch response. In this circuit, resistances $R1$ and $R2$ govern the gain at frequencies above and below that of the notch. Resistances $R3$ and $R4$ should equal each other; let's call that resistance R. In addition, capacitances $C1$ and $C2$ should equal each other; let's call that capacitance C.

The resonant notch (or minimum gain) frequency drops if you:

- Increase R but leave C constant.
- Increase C but leave R constant.
- Increase both R and C.

The resonant notch frequency rises if you:

- Decrease R but leave C constant.
- Decrease C but leave R constant.
- Decrease both R and C.

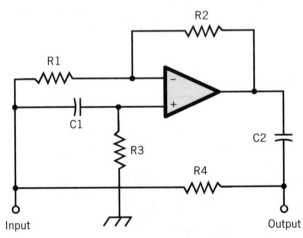

FIG. 5-25 *An op amp wired up to serve as a resonant notch filter. The values of resistors and capacitors determine the resonant frequency. In this circuit, R3 = R4; you can call this resistance R. Also, C1 = C2; you can call this capacitance C.*

To calculate the resonant notch frequency f (in hertz), you need to know R and C, as defined above, in ohms and farads respectively. Then you can use the formula

$$f = 1 \; / \; (2\pi RC)$$

This formula will also yield f in hertz if you express both $R3$ and $R4$ in megohms and both $C3$ and $C4$ in microfarads.

Summary

Reading and drawing schematics often involve breaking down complex circuits into simple ones. Then you can look at the system's parts and how they relate to each other, rather than try to imagine the whole thing as a single appliance. As you study a complex schematic, the relationships among the circuits will grow apparent. Once in awhile, you'll see all of a system's "secrets" revealed at once: an "Aha" moment!

> **Tip**
> When you use schematics for electronics troubleshooting, you don't necessarily have to understand the function of every single system element. In many cases, you'll only have to concern yourself with those circuits and/or components that represent potential trouble spots.

Learning to read and write schematics is a lot like learning to receive and send the old Morse code. "The code" is a language of audible symbols, just like a schematic is a language of printed symbols. Once you learn either language, you can use it to communicate; Morse code communicates words and sentences, while schematics communicate principles and concepts.

Using Morse code as a further example, a long sequence of dots and dashes will mean nothing unless you can break the data down into words. As your proficiency increases, you'll stop hearing the individual dots and dashes (or, as some people say, "dits" and "dahs") and hear letters of the alphabet instead. As you keep practicing, you'll start to hear entire words. Eventually, if you keep at it long enough (and especially if you get fond of the code for its own sake, communicating with it for hours on end, as I have done over the years as a ham radio operator), you'll hear whole phrases and sentences.

Reading and writing schematics takes you through a similar pattern of development. At first you'll see individual component symbols. Later, you'll find simple circuits hidden within complex circuits. Then you'll identify and analyze those complex circuits. Finally, you'll envision entire systems. This knowledge might come slowly, but your proficiency will improve every time you practice if you keep pushing yourself (gently, of course) into new knowledge zones.

6

Diagrams for Building and Testing

Several years ago, McGraw-Hill published my book *Electricity Experiments You Can Do at Home*. This chapter contains a few experiments from that book along with drawings and diagrams. These experiments can help you gain proficiency in reading and interpreting schematics, pictorials, and real-world circuits. Let's start with some setup details including a parts list and the construction of a circuit board called a *breadboard*. Then you can try these experiments if you like. If you enjoy them, you can get a copy of *Electricity Experiments You Can Do at Home* and do more!

Tip
You needn't buy all the parts, build the breadboard, and do all the experiments described here. You'll have fun and learn things better that way, but if you don't want to spend money and time on "nuts and bolts," you can learn a fair amount by following along and conducting the experiments in your imagination.

Your Breadboard

Every experimenter needs a good workbench. Mine comprises a piece of plywood, weighted down over the keyboard of an old upright piano, and hung from the cellar ceiling by brass-plated chains. Yours doesn't have to be that exotic. You can put it anywhere as long as it won't shake or collapse. You should make the surface out of a solid non-conducting material such as wood, protected by a plastic mat or a small piece of close-cropped carpet (a door-mat works great). A desk lamp, preferably the "high-intensity" type with an adjustable arm, completes the ensemble.

Before you begin any of the tasks described in this chapter, buy a pair of safety glasses at your local hardware store. Wear the glasses whenever you build and test these circuits. Get into the habit of wearing safety glasses whether you think you need them or not. You never know when a little piece of wire will fly at one of your eyes as you snip it off with diagonal cutter!

Bits and Pieces

Table 6-1 lists the items you'll need for the experiments in this chapter. In the "good old days" you could find components at many Radio Shack retail stores. You can still get parts online from them (and from some remaining retail stores). Otherwise, you can take advantage of electronics parts suppliers such as those listed in Appendix C at the back of this book. You'll also find some of these items at hardware or department stores.

Your breadboard doesn't have to look fancy or cost much money. I patronized a local lumber yard to get the wood for mine. I found a length of "12-inch by 3/4-inch" pine in their scrap heap. The actual width of a "12-inch" board is about 10.8 inches (27.4 centimeters), and the actual thickness is about 0.6 inch (15 millimeters). The people at the lumber yard didn't charge me anything for the wood itself, but they demanded a couple of dollars to make a clean cut to produce a rectangular piece of pine measuring 12.5 inches (31.8 centimeters) long.

Using a ruler, divide the breadboard lengthwise at 1-inch (25.4-millimeter) intervals, centered to get 11 evenly spaced marks. Do the same thing going sideways to obtain nine marks at 1-inch (25.4-millimeter) intervals. Using a ball-point or roller-point pen, draw lines parallel to the edges of the board to make a grid pattern. Label the grid lines from A to K and 0 to 9 as

TABLE 6-1 *Components list for electricity experiments. You can find these items at retail stores or order them from vendors such as those in Appendix C at the back of this book.*
Abbreviations: AWG = American Wire Gauge, A = amperes, V = volts, W = watts, and PIV = peak inverse volts. In some cases this table includes more items than you'll likely need.
(You never know when you'll want an "extra" or two!)

Quantity	Store type	Description
1	Lumber yard	Pine board, approx. 10.8 × 12.5 × 0.6 inches
1	Hardware store	Pair of safety glasses
1	Hardware store	Small hammer
24	Hardware store	Flat-head wood screws, 6 × 3/4
150	Hardware store	Polished steel finishing nails, 1-1/4 inches long
4	Hardware store	Sheet of fine-grain sandpaper
1	Stationery store or department store	Standard paper punch for making 1/4-inch holes
1	Department store	12-inch plastic or wooden ruler
1	Hardware store	Small tube of contact cement
1	Hardware store or online vendor	Digital multimeter (volt-ohm-milliammeter)
1	Hardware store	Diagonal wire cutter/stripper
1	Hardware store	Small needle-nose pliers
2	Hardware store	Roll of AWG No. 24 solid uninsulated (bare) copper wire
1	Hardware store	Small roll of thin enameled "magnet wire"
15 to 20	Online vendor	Insulated test/jumper leads
1	Sporting goods store, hardware store, or department store	Hiker's "pocket compass" calibrated in degrees
12	Hardware store	Alkaline AA cells rated at 1.5 V
2	Online vendor	Holder for four size AA cells in series
4	Online vendor	Holder for one AA cell
10	Online vendor	Resistor rated at 220 ohms and 1/2 W
10	Online vendor	Resistor rated at 330 ohms and 1/2 W
10	Online vendor	Resistor rated at 470 ohms and 1/2 W
10	Online vendor	Resistor rated at 680 ohms and 1/2 W
10	Online vendor	Resistor rated at 1000 (1 k) ohms and 1/2 W
10	Online vendor	Resistor rated at 1500 (1.5 k) ohms and 1/2 W

(continued on next page)

TABLE 6-1 *(Continued) Components list for electricity experiments. You can find these items at retail stores or order them from vendors such as those in Appendix C at the back of this book.* Abbreviations: AWG = American Wire Gauge, A = amperes, V = volts, W = watts, and PIV = peak inverse volts. In some cases this table includes more items than you'll likely need. (You never know when you'll want an "extra" or two!)

Quantity	Store type	Description
10	Online vendor	Resistor rated at 3300 (3.3 k) ohms and 1/2 W
6	Online vendor	Rectifier diode rated at 1 A and 600 PIV
4	Online vendor	Screw-base miniature lamp holder
4	Online vendor	Screw-base miniature lamp rated at 6.3 V DC
4	Online vendor	Screw-base miniature lamp rated at 7.5 V DC

shown in Fig. 6-1. That'll give you 99 intersection points, each of which you can designate with a letter/number pair such as D-3 or G-8.

Once you've marked the grid lines, gather together a bunch of 1.25-inch (31.8-millimeter) polished steel finishing nails. Place the board on a solid surface that can't suffer any damage from scratching or scraping. Pound a nail into each intersection point shown by the black dots in Fig. 6-1. Make certain that the nails have "tiny heads" and *do not* have any coating of paint, plastic, lacquer, or other electrically insulating material. Each nail should go into the board just far enough so that you can't wiggle it around. I pounded every nail down to a depth of approximately 0.3 inch (8 millimeters), a distance amounting to halfway through the board.

Use some 6 × 32 flat-head wood screws to secure the two miniature lamp holders to the board at the locations shown in Fig. 6-1. Connect the terminals of one lamp holder to breadboard nails A-2 and D-1 with short lengths of thin, solid, bare copper wire. Connect the terminals of the other lamp holder to nails D-2 and G-1. Wrap the wire tightly three or four times around each nail. Snip off any excess wire that remains. Glue the four-cell AA battery holder to the breadboard with contact cement. Allow the cement to harden. That process will need a few hours, so you can take a break for awhile!

When the contact cement has solidified, strip approximately 1 inch (25.4 millimeters) of the insulation from the ends of the cell-holder leads and connect the leads to the nails as shown in Fig. 6-1. Remember that the red lead goes to the positive battery terminal, and the black lead goes to the negative terminal. Use the same wire-wrapping technique that you did for the lamp-holder wires. Place four brand new AA alkaline cells in the holders with the

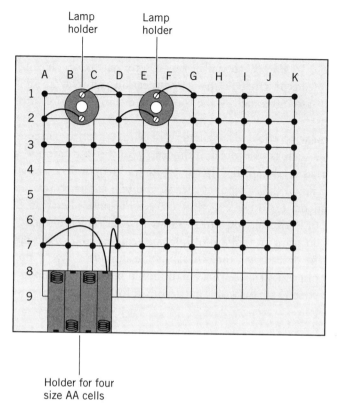

FIG. 6-1 *Breadboard layout for electricity experiments. Solid dots show nail locations. Grid squares measure 1 by 1 inch (25.4 by 25.4 millimeters).*

negative terminals against the springs. I recommend that you buy a single package of four cells so that they're all equally "fresh." Now you have a 6-V battery, and the breadboard awaits your exploits.

Go Ahead and Laugh!
When the previous edition of this book hit the shelves, some readers made fun of this "homebrew" breadboard. It's sort of rustic, I admit, but you can always find the parts to make it, and it always works. So go ahead and laugh if you must, and then get on with the experiments!

Wire Wrapping

The following breadboard-based experiments employ a construction method called *wire wrapping*. Each nail forms a terminal to which you can attach component leads or wires. To make a connection, tightly wrap an uninsulated wire or lead around a nail. Make four or five complete wire turns as shown in Fig. 6-2.

After you wrap the end of a length of wire, cut off the excess. For small components such as resistors and diodes, wrap the leads around the nails as many times as it takes to use up the entire lead length. That way, you won't have to cut down the component leads, and you can later unwrap and reuse the components for other experiments. Needle-nose pliers can help you to wrap wires or leads that you can't wrap with your fingers alone.

When you want to make multiple connections to a single nail, you can wrap one wire or lead over the other, but you shouldn't have to do that unless you've run out of nail space. Each nail should protrude far enough above the board surface so that you won't get cramped for wrapping space very often.

Tip

Again, you must make certain that the nails comprise steel *without any coating*. They should be new and clean, so they'll function as efficient electrical terminals. If necessary, you can use fine-grain sandpaper or an emery board to guarantee that you'll get good connections.

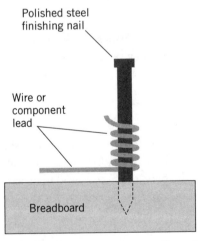

Polished steel finishing nail

Wire or component lead

Breadboard

FIG. 6-2 *Wire-wrapping technique. If necessary, you can use a diagonal cutter to snip off excess wire.*

As you build the circuits that follow, you can tailor the arrangement of parts on the breadboard to suit your needs. I've provided schematic and pictorial diagrams to show you how the components interconnect. I recommend that you follow my layout suggestions. That way, you can focus on how the actual appearance of the circuit compares with the schematic, even if you haven't bothered to build the breadboard and work with the hardware directly.

Small components such as resistors should go between adjacent nails (either straight or diagonally) so you can wrap each lead securely around each nail. If the nails lie too distant from each other, the component leads might not reach far enough to allow decent wrappings. You should secure *jumper wires*, also known as *clip leads*, to the nails so that their "jaws" can't easily work their way loose. I suggest that you clamp the jumpers to the nails sideways so the wires come off horizontally. If you try to put one of these so-called *alligator clips* down on a nail vertically, it might pop off in the middle of a mission-critical operation!

Caution!

Please let me repeat: Wear safety glasses at all times as you perform these experiments, whether you think you need them or not.

Kirchhoff's Current Law

In this experiment, you'll construct a network that demonstrates one of the most important principles in DC electricity. You'll need five resistors: two rated at 330 ohms, one rated at 1000 ohms (1 k), and two rated at 1500 ohms (1.5 k). You'll also need four AA cells. In my opinion, Duracell and Eveready sell the best electrochemical cells and batteries in the United States.

Mount the resistors on the breadboard by connecting them between pairs of terminals as shown in Fig. 6-3. Test each resistor with your multimeter (set to function as an ohmmeter) to verify its ohmic value before you install it. Use a 5-inch (13-centimeter) length of bare copper wire to interconnect the three terminals I-1, J-1, and K-1. Do the same thing with I-3, J-3, and K-3, and also with I-5, J-5, and K-5.

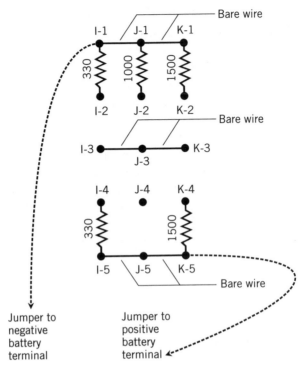

FIG. 6-3 *Arrangement of resistors on breadboard for demonstration of Kirchhoff's current law. All resistance values are in ohms. Solid dots indicate breadboard terminals. Solid lines show interconnections with bare copper wire. Dashed lines indicate jumpers.*

Tip

Engineers symbolize current *as a variable* by writing an uppercase italic letter *I*. Amperes *as units* get abbreviated as an uppercase non-italic A. Voltage *as a variable* can be symbolized by an uppercase italic *E* or an uppercase italic *V*. Volts *as units* get abbreviated as an uppercase non-italic V. Resistance *as a variable* is symbolized with the uppercase italic *R*. Ohms *as units* can be written out in full (ohm or ohms), although some texts use the uppercase non-italic Greek letter omega (Ω).

Gustav Robert Kirchhoff (1824–1887) did research and formulated theories in a time when people didn't know much about electrical current. He used common sense to deduce the fundamental properties of DC circuits. Kirchhoff reasoned that the current entering any branch point in a circuit must always equal the current leaving that point. *Kirchhoff's current law* holds true no matter how many branches enter a given point, and no matter how many branches leave it.

Follow the Flow

Figure 6-4 shows a generic example of Kirchhoff's current law. You can also call it *Kirchhoff's first law* or the *principle of current conservation*. In Fig. 6-4, two resistors enter the branch point and three resistors leave it, so

$$I_1 + I_2 = I_3 + I_4 + I_5$$

Connect your four-cell battery to the resistive network and measure the currents in each branch. Meter each test point individually while all the other test points remain shorted with jumpers. The schematic of Figure 6-5 shows the actual values of the resistors in my network (yours will doubtless differ slightly from mine), along with the value I got when I measured I_1, the current through the smaller of the two input resistors.

As you measure each current value, make sure that the meter polarity agrees with the battery polarity. The black probe should go to the more negative point, and the red probe should go to the more positive point. That way, you'll avoid negative current readings that might throw off your calculations.

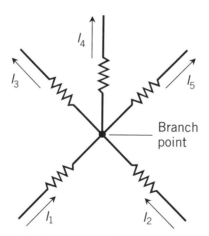

FIG. 6-4 *According to Kirchhoff's current law, the sum of the currents flowing into any branch point always equals the sum of the currents flowing out of that branch point. In this example, $I_1 + I_2 = I_3 + I_4 + I_5$.*

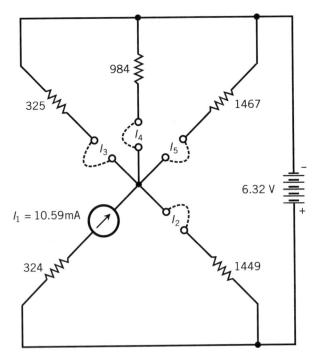

FIG. 6-5   *Network for verifying Kirchhoff's current law. All resistance values are in ohms. The battery voltage, the current I_1, and the resistances indicate my measurements. Dashed lines show interconnections with jumpers.*

When I tested my four-cell battery to determine its voltage, I got 6.32 V. When I measured I_1 through I_5, I got the following results, accurate to the nearest hundredth of a milliampere (mA):

$$I_1 = 10.59 \text{ mA}$$
$$I_2 = 2.40 \text{ mA}$$
$$I_3 = 8.35 \text{ mA}$$
$$I_4 = 2.79 \text{ mA}$$
$$I_5 = 1.88 \text{ mA}$$

Again, make certain that every pair of test points *not* undergoing current measurement gets shorted together with jumpers. Otherwise, you'll have an incomplete network, and your current measurements will turn out wrong. After you've finished making the measurements, remove all the jumpers to conserve battery energy.

Now you can input your numbers to the Kirchhoff formulas and see how close the sum of the input currents comes to the sum of the output currents. Here are my results for the sum of the currents entering the branch point:

$$I_1 + I_2 = 10.59 + 2.40$$
$$= 12.99 \text{ mA}$$

When I added the currents leaving the branch point, I got

$$I_3 + I_4 + I_5 = 8.35 + 2.79 + 1.88$$
$$= 13.02 \text{ mA}$$

Tip

When you do experiments of this sort, you should expect a slight discrepancy. That "uncertainty principle" explains the 0.03 mA current difference in the branch points during my test. An error of three hundredths of a milliampere at 14 mA amounts to well under one percent, which is acceptable.

Follow the Flow

Compare the layout (Fig. 6-3) with the schematic (Fig. 6-5) and follow the current as it flows through the resistors in both diagrams. Note that in Fig. 6-3, the resistors are labeled with their *rated* values, while Fig. 6-5 shows the *actual* values as I measured them.

Kirchhoff's Voltage Law

In this experiment, you'll construct a network that demonstrates another important DC circuit rule. You'll need four resistors: one rated at 220 ohms, one rated at 330 ohms, one rated at 470 ohms, and one rated at 680 ohms. You'll need four AA cells again, too.

According to *Kirchhoff's voltage law*, the sum of the potentials (voltages) across the individual components in a series DC circuit, taking polarity into account, always equals zero. You can also call this rule *Kirchhoff's second law* or the *principle of voltage conservation*.

Ponder the Potentials

Figure 6-6 shows a generic example of Kirchhoff's voltage law. The battery potential, E, equals the sum of the potentials across the resistors. The polarity across each resistor opposes the polarity of the battery, so that:

$$E + E_1 + E_2 + E_3 + E_4 = 0$$

If you measure the potentials across the individual resistors and the battery, one at a time, with a DC voltmeter and *disregard the polarity*, you'll find that:

$$E = E_1 + E_2 + E_3 + E_4$$

Check each of the four resistors with your ohmmeter to verify their values before you install them in the circuit. Mount the resistors in the upper right-hand corner of your breadboard by wrapping the leads around the nails to get the arrangement of Fig. 6-7. Connect the battery to the network with jumpers as indicated, and then measure the voltage across each resistor. Figure 6-8 illustrates the actual values of the resistors in my network (no doubt yours will differ a bit), along with the resistance across which E_2 appeared. I measured $E = 6.30$ V across the battery when I connected it to the resistors.

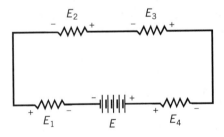

FIG. 6-6   *According to Kirchhoff's voltage law, the sum of the voltages across the resistances in a series DC circuit always equals the battery voltage, but with the opposite polarity. If you disregard polarity, then in this example you'll get $E = E_1 + E_2 + E_3 + E_4$.*

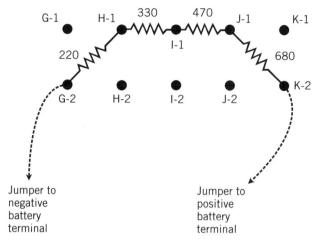

FIG. 6-7 *Suggested arrangement of resistors on breadboard for demonstration of Kirchhoff's voltage law. All resistance values are in ohms. Solid dots indicate terminals. Dashed lines indicate jumpers.*

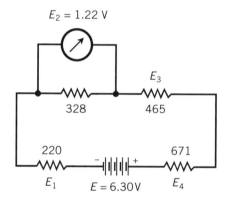

FIG. 6-8 *Network for verifying Kirchhoff's voltage law. All resistances are in ohms. The battery voltage E, the voltage across the second resistor, and the resistances are the values I measured.*

As you measure each voltage E_1 through E_4, the black meter probe should go to the more negative voltage point and the red probe should go to the more positive point to avoid negative readings that could mess up your calculations. When I measured the voltages across the individual resistors, I got

$$E_1 = 0.82 \text{ V}$$
$$E_2 = 1.22 \text{ V}$$
$$E_3 = 1.75 \text{ V}$$
$$E_4 = 2.52 \text{ V}$$

When you finish making your measurements, remove one of the jumpers to take the stress off the battery.

After you've double-checked and written down your voltage measurements, input the numbers to the modified Kirchhoff formula:

$$E = E_1 + E_2 + E_3 + E_4$$

and see how closely it works out. I got the following results:

$$E = 6.30 \text{ V}$$

and:

$$E_1 + E_2 + E_3 + E_4 = 0.82 + 1.22 + 1.75 + 2.52$$
$$= 6.31 \text{ V}$$

That's an error of only 0.01 V out of a total potential of 6.30 V, amounting to less than two-tenths of one percent.

Follow the Flow

Compare the layout (Fig. 6-7) with the schematic (Fig. 6-8). The resistors are connected in series. Therefore, the same current flows through each and every one of them. The voltage E_x across any given resistance R_x depends directly on the current I_x through the component according to Ohm's law:

$$E_x = I_x R_x$$

Again, Fig. 6-7 shows the rated resistances, while Fig. 6-8 shows the ohmic values as I measured them.

A Resistive Voltage Divider

You can use the components from the previous experiment to get several different voltages from a single battery. Keep the resistors on the breadboard in the same arrangement as you had them in the experiment for Kirchhoff's voltage law.

When you connect two or more resistors in series with a DC power source, those resistors produce various voltage ratios. You can tailor these ratios using specific resistances that "fix" the intermediate voltages. The circuit works best when your network resistors have values much smaller than the resistance of any external load that you place across the combination.

Figure 6-9 illustrates the principle of a *resistive voltage divider*. The individual resistances are R_1, R_2, R_3, ..., and R_n. The total resistance R equals their sum:

$$R = R_1 + R_2 + R_3 + ... + R_n$$

Follow the Flow

If you call the power supply voltage E, then Ohm's law tells you the current I at all points in the series circuit of Fig. 6-9:

$$I = E / R$$

as long as you express I in amperes, E in volts, and R in ohms.

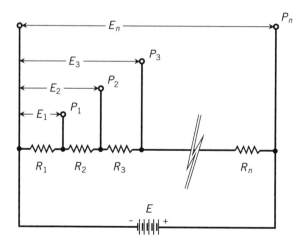

FIG. 6-9 *A voltage divider takes advantage of the potential differences across individual resistors connected in series with a DC power source. Note the use of italics for voltages, test points, and resistances.*

At points P_1, P_2, P_3, ..., and P_n, let's call the voltages relative to the negative battery terminal E_1, E_2, E_3, ..., and E_n, respectively. The last (and highest) voltage, E_n, equals the battery voltage, E. The voltages at the various points increase according to the sum total of the resistances up to each point, in proportion to the total resistance, multiplied by the battery voltage. In theory, the following equations hold true:

$$E_1 = E R_1 / R$$
$$E_2 = E (R_1 + R_2) / R$$
$$E_3 = E (R_1 + R_2 + R_3) / R$$

$$\downarrow$$

$$E_n = E (R_1 + R_2 + R_3 + ... + R_n) / R = E R / R$$
$$= E$$

During this experiment, I measured $E = 6.30$ V across the battery when it operated under a load of all the series-connected resistors in Figs. 6-10A and B. Then I connected the voltmeter across the series combination of the first two resistors only. (Figure 6-10A shows the rated values of the resistors while Fig. 6-10B shows the values I read from my ohmmeter.) I measured the resistances as follows:

$$R_1 = 220 \text{ ohms}$$
$$R_2 = 328 \text{ ohms}$$
$$R_3 = 465 \text{ ohms}$$
$$R_4 = 671 \text{ ohms}$$

Set your meter to measure current in milliamperes (mA). Connect the battery to the resistive network through the meter as shown in Fig. 6-10B, and measure the current. According to strict theory, I expected the milliammeter to indicate a value equal to the battery voltage divided by the sum of my measured resistances:

$$I = E / R$$
$$= 6.30 / (220 + 328 + 465 + 671)$$
$$= 6.30 / 1684$$
$$= 0.00374 \text{ A}$$
$$= 3.74 \text{ mA}$$

When I measured the current, I got 3.73 mA, a value only a hundredth of a milliampere (0.01 mA) different from the theoretical value!

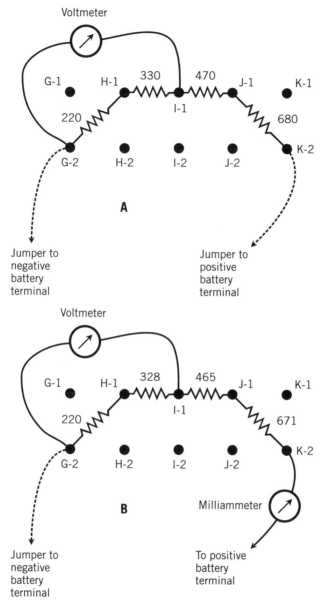

FIG. 6-10 *At A, arrangement for measuring voltages in a resistive divider. Here, the voltmeter is connected to measure the potential difference E_2 across the series combination of the first and second resistors. All resistances are in ohms. Solid dots indicate terminals. Dashed lines indicate jumpers. At B, arrangement for measuring current through the network. At A, the resistances are the rated values. At B, the resistances are my measured values.*

Now measure the intermediate voltages E_1 through E_4 with your meter set for a moderate DC voltage range. The black meter probe should go directly to the negative battery terminal and stay there. The red meter probe should go to each positive voltage point in turn. First measure the voltage E_1 that appears across R_1 only. Then measure, in order, the voltages as illustrated in the schematic of Fig. 6-9 (assuming $n = 4$):

- The potential E_2 across $R_1 + R_2$
- The potential E_3 across $R_1 + R_2 + R_3$
- The potential E_4 across $R_1 + R_2 + R_3 + R_4$

Ponder the Potentials

Figure 6-11 illustrates the arrangement for measuring E_2 to provide a specific example. (It also shows the milliammeter connection as Fig. 6-10B does.) Again, this schematic shows the resistances as I measured them. When I tested all four voltages, my meter displayed these results:

$$E_1 = 0.82 \text{ V}$$
$$E_2 = 2.04 \text{ V}$$
$$E_3 = 3.79 \text{ V}$$
$$E_4 = 6.30 \text{ V}$$

After you've finished measuring the voltages, remove one of the jumpers to conserve battery energy.

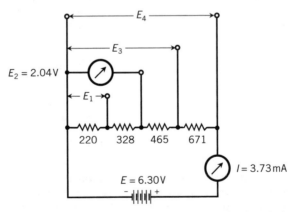

FIG. 6-11 *Circuit for testing the operation of a resistive voltage divider. The battery voltage E, the potential difference E_2 across the series combination of the first and second resistors, and the resistances represent my measurements.*

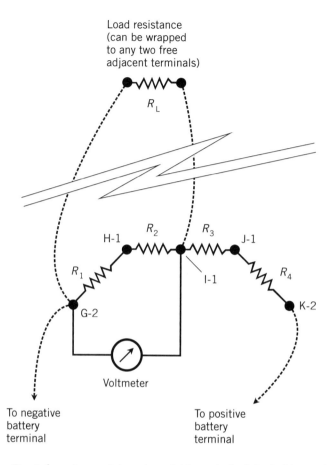

FIG. 6-12 *Circuit for testing a resistive voltage divider under load. Dashed lines indicate jumpers. This diagram shows the arrangement for measuring variations in E_2 as the load resistance R_L is alternately connected and disconnected from the series combination of R_1 and R_2.*

Now connect the meter across the combination $R_1 + R_2$. Run jumper wires to the ends of a *load resistor* located elsewhere on the breadboard, as shown in the layout diagram of Fig. 6-12. This arrangement will cause the voltage source E_2 to drive some current through the load resistor (which we'll call R_L), in addition to some current that will keep flowing through R_1 and R_2. Try every resistor in your repertoire in the place of R_L. If you obtained all the resistors in the parts list, you'll have seven tests to do, using resistors rated at values ranging from 220 to 3300 ohms.

Alternately, connect and disconnect one of the jumper wires between the voltage divider and R_L, so you can observe the effect of the extra load on E_2. As you'll see, the additional load affects the behavior of the voltage divider. As R_L decreases, so does E_2. The effect grows more dramatic as R_L decreases, representing a "heavier and heavier load." Table 6-2 shows the results I got.

Plot your results as points on a coordinate grid with R_L on the horizontal axis and E_2 on the vertical axis. Connect the dots to approximate a *characteristic curve* that shows E_2 as a function of R_L. Figure 6-13 is the graph I made. I used a *reverse logarithmic scale* to portray R_L so the values would appear spread out. This graph provides a clear picture of what happens as the *conductance* of the load *increases*.

Here's a Thought!

What do you think will happen to the voltage across the load if you use two, three, four, or five 220-ohm resistors in parallel, getting R_L values of about 110, 73, 55, and 44 ohms, respectively? Try the arrangements one by one and see what happens! (You originally bought more than enough resistors, didn't you?)

TABLE 6-2 *Here are the voltages that I measured across various loads in a resistive voltage divider constructed according to Fig. 6-12. My measured network resistances, indicated in the text, were R_1 = 220 ohms, R_2 = 328 ohms, R_3 = 465 ohms, and R_4 = 671 ohms. My measured load resistances (left column), which differed a bit from the network resistances (because they involved different physical components!), were the actual values for resistors rated at 3300, 1500, 1000, 680, 470, 330, and 220 ohms, respectively as you read down.*

Load resistance (ohms)	Load voltage (volts)
3250	1.83
1470	1.63
983	1.49
672	1.32
466	1.14
326	0.96
220	0.76

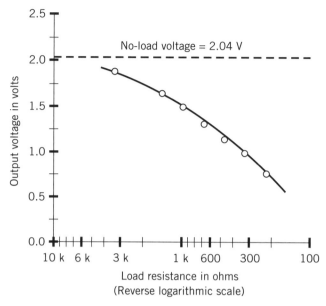

FIG. 6-13   *My measurements of output voltage vs. load resistance R_L in the voltage divider. The dashed line shows the open-circuit (no-load) voltage across the combination of R_1 and R_2 in series. Open circles show the measured voltages across various loads. The solid curve lets you see how the circuit behaves as R_L goes down.*

Beware!

The results of this experiment suggest that when engineers build voltage dividers, they had better know what sort of external load the circuit will have to contend with. If the load resistance fluctuates greatly, especially if it sometimes gets low, a resistive voltage divider won't perform well. In the worst case, it'll prove useless.

A Diode-Based Voltage Reducer

Rectifier diodes can reduce the output voltage of a DC power source, providing a more predictable way to get specific voltages than a resistive divider can do. Let's build one! For this experiment, you'll need two rectifier diodes. Those that I obtained were rated at 1 A and 600 *peak inverse volts* (PIV), although higher ratings will work too. You'll also need at least one of each resistor listed in Table 6-1, along with some jumpers.

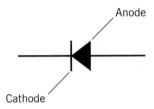

FIG. 6-14   *Schematic symbol for a semiconductor diode. The line represents the cathode. The arrow represents the anode.*

Figure 6-14 shows the schematic symbol for a rectifier diode. Manufacturers can produce one of these things by joining a piece of *P-type* semiconductor material to a piece of *N-type* material, creating a so-called *P-N junction*. The N-type semiconductor, represented by the short, straight line, forms the *cathode*. The P-type semiconductor, represented by the arrow, forms the *anode*. Under most conditions, electrons can move easily from the cathode to the anode (against the arrow), but not from the anode to the cathode (with the arrow). Conventional or theoretical current, which always goes from positive to negative, flows in the direction that the arrow points.

Follow the Flow

If you connect a battery and a resistor in series with a diode, current flows if the negative battery terminal faces the cathode and the positive terminal faces the anode (Fig. 6-15A). This condition is called *forward bias*. No current flows if you reverse the battery (Fig. 6-15B) unless the voltage gets extreme. This condition is called reverse bias. The series resistor protects the diode from excessive current as long as the battery voltage remains reasonable.

It takes a certain minimum voltage to drive current through a forward-biased diode. Engineers call this threshold the *forward breaker voltage*. In most diodes it's a fraction of a volt, but it varies somewhat depending on how much current you force the diode to carry. If the forward-bias voltage across the P-N junction does not equal or exceed the forward breaker voltage, then the diode won't conduct. When you forward-bias a diode and connect it in series with a resistor and a battery, the P-N junction voltage goes down to an extent approximately equal to the forward breakover voltage. Unlike the voltage reduction that takes place with resistors, a diode's voltage-dropping capability stays nearly constant as you vary the external load resistance.

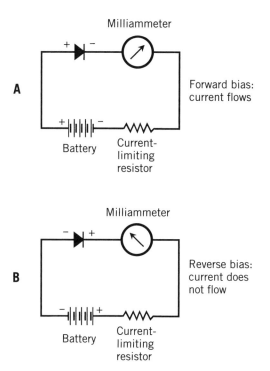

FIG. 6-15   *Series connection of a battery, resistor, current meter, and diode. At A, forward bias allows current to flow if the voltage equals or exceeds the forward breakover threshold. At B, reverse bias drives no current through the diode (unless the voltage gets very high).*

Although current won't normally flow through a reverse-biased diode, exceptions occur. If the reverse voltage gets high enough (usually much more than the forward breakover value), a diode will conduct because of so-called *avalanche effect*. Zener diodes, often used to regulate DC power-supply voltage, work according to this principle.

> **Tip**
> When you connect two or more identical rectifier diodes in series with their polarities in agreement, the forward-breakover voltages of each diode add (give or take a little), so you can build circuits to produce stable, predictable voltage drops that come close to whole-number multiples of the forward breakover voltage of a single diode. This technique works only if you connect an external load to the circuit, forcing the diodes to carry some current.

You can set up a voltage reducer with two diodes in series and their polarities in agreement as shown in Fig. 6-16, so that current flows through the load resistor R_L if the forward bias voltage is high enough. Figure 6-17 shows an arrangement for mounting the components on your breadboard. Set your meter to indicate DC voltage in a moderate range such as 0 to 20 V. Connect the meter across the load resistance R_L, paying attention to the polarity so that you'll get positive meter readings. Try each one of your resistors in the place of R_L. Measure the voltage across R_L in every case. You'll have seven tests to do, with rated resistances ranging from 220 to 3300 ohms.

The load resistance R_L affects the behavior of a diode-based voltage reducer, but in a different way than it affects the behavior of a resistive voltage divider. As you perform these tests, you'll see that as the load resistance R_L decreases, the potential difference across it goes down, but only a little. The voltage across the load tends to drop *more and more slowly* as R_L decreases. Contrast this behavior with that of the resistive voltage divider, in which the voltage drops off *more and more rapidly* as R_L decreases. Table 6-3 shows the results I got when I measured the voltages across various values of R_L with this two-diode arrangement.

Plot your results as points on a coordinate grid with the load resistance R_L on the horizontal axis and the voltage across R_L on the vertical axis, and then approximate the curve as you did in the previous experiment. I got the graph of Fig. 6-18. As before, I used a reverse logarithmic scale to portray the load resistance. Compare this graph with Fig. 6-13 from the previous experiment.

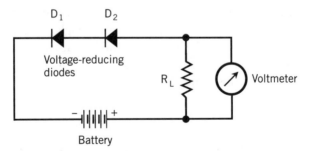

FIG. 6-16 *Schematic showing voltage measurement across the load resistance in a two-diode voltage reducer.*

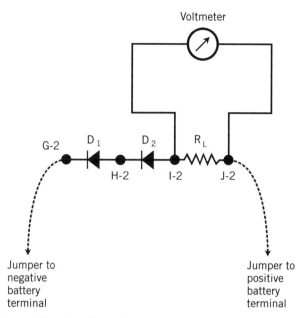

FIG. 6-17 *Suggested breadboard layout for measuring voltage across the load resistance R_L in a two-diode voltage reducer. Solid dots show breadboard terminals. Dashed lines indicate jumpers. Pay attention to the diode polarities! The cathodes should face toward the negative battery terminal.*

TABLE 6-3 *Here are the output voltages that I obtained with various loads connected to a diode-based voltage reducer. These are the resistances that I measured. (Yours will of course differ a bit.) The circuit comprised two diodes rated at 1 A and 600 PIV, forward-biased and connected in series with a 6.30-V battery.*

Load resistance (ohms)	Load voltage (volts)
3250	5.08
1470	4.99
983	4.96
671	4.91
466	4.88
326	4.84
220	4.79

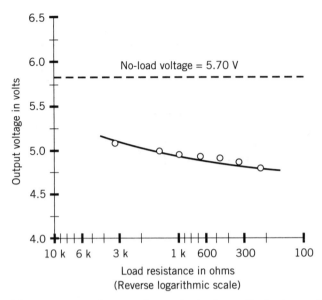

FIG. 6-18   *My measurements of output voltage vs. load resistance R$_L$ for the two-diode voltage reducer. The dashed line shows the open-circuit (no-load) voltage. Open circles show measured voltages across various loads. The solid curve reveals how the circuit behaves as R$_L$ goes down.*

Here's a Thought!

Repeat this experiment with only one diode. Then try the experiment with three, four, or five diodes in series. You might also obtain some more resistors, covering a range of values of, say, 100 ohms to 100,000 ohms, and test the circuit using them for R$_L$.

Caution!

Don't use a resistor of less than about 75 ohms as the load here. In this arrangement, a 1/2-watt resistor of less than 75 ohms will allow excessive current to flow, risking destruction of the resistor (and maybe the diodes too!).

Mismatched Lamps in Series

When two dissimilar incandescent lamps operate in series, they receive different voltages and consume different amounts of *volt-ampere* (VA) power, as this experiment demonstrates. (Remember from your basic electricity courses that in a DC circuit, power in watts equals voltage in volts times current in amperes, hence the term *volt-ampere* for simple DC power.) You'll need a 6.3-V lamp, a 7.5-V lamp, a battery of four AA cells, and your multimeter set to measure voltage, as shown in the schematic of Fig. 6-19. You'll also need some jumpers.

Your breadboard should have two screw-base lamp holders. Position the board so that both holders lie near the top, side by side (Fig. 6-20). Install a 6.3-V lamp in the left-hand socket, and a 7.5-V lamp in the right-hand socket. Connect a short length of bare wire securely between terminals D1 and D2, so the top terminal of the left-hand lamp holder goes to the bottom terminal of the right-hand lamp holder. Then connect jumpers between the free lamp socket terminals and the battery terminals so the lamps operate in series. When you connect the battery to send current through the lamps, they both should glow at partial brilliance.

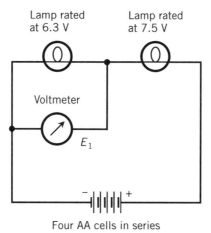

FIG. 6-19 *Measurement of voltage E_1 across the more negative of two dissimilar lamps in series.*

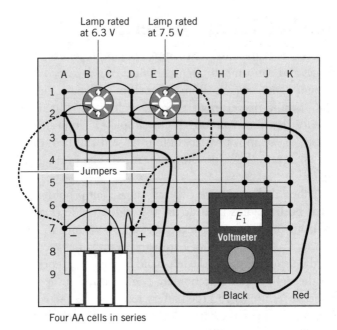

FIG. 6-20   *Breadboard layout for the schematic of Fig. 6-19.*

Call the lamp closer to the negative battery terminal "lamp N." That should be the one on the left. Call the lamp closer to the positive battery terminal "lamp P." That should be the one on the right. Unscrew lamp N. When contact fails, both lamps will go dark. Screw lamp N back in, and then unscrew lamp P. Once again, both lamps will go dark. This demonstrates how a series circuit behaves: A break at any point prevents current from flowing anywhere in the circuit.

Now short out lamp N (the one rated at 6.3 V) with a jumper. It will go dark because it no longer has any voltage across it, while lamp P (rated at 7.5 V) will attain nearly full brilliance. Then disconnect the jumper from lamp N and move the jumper to short out lamp P instead. Lamp P will go dark while lamp N glows at 100 percent of full brilliance. Again, this behavior typifies series circuits. If any component shorts out, all the others consume more power than before.

Set your meter to read DC volts. Connect the jumpers so the lamps are in series and both glow at partial brilliance. Measure the voltage E_1 across lamp N as shown in Figs. 6-19 and 6-20. Then measure the voltage E_2 across lamp P as shown in the schematic of Fig. 6-21 and the equivalent layout pictorial of Fig. 6-22. When I performed these tests, I got the meter readings:

$$E_1 = 2.20 \text{ V}$$

and:

$$E_2 = 3.64 \text{ V}$$

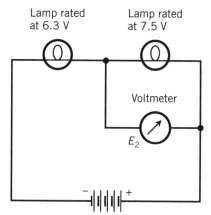

FIG. 6-21 *Measurement of voltage across the more positive of two dissimilar lamps in series.*

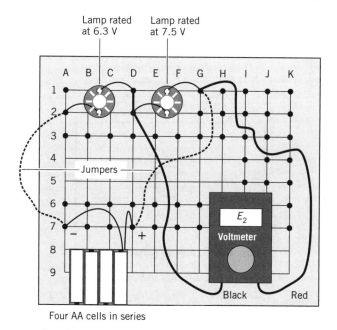

FIG. 6-22 *Breadboard layout for the schematic of Fig. 6-21.*

Tip

If you can't get the exact same lamps as specified in the parts list for these experiments (Table 6-1), you can use other lamps, within certain limits. This experiment will work as long as the lamps are rated for slightly different voltages or power levels, and both are designed to work with DC at voltages between 6 V and 12 V.

Now determine the voltage E across the series combination of lamps as shown in the schematic of Fig. 6-23 and the layout pictorial of Fig. 6-24. In theory, your meter should display the sum of the lamp voltages, or $E = E_1 + E_2$, which equals the battery voltage. When I input the results from my individual lamp tests into this formula, I predicted that I would see

$$E = 2.20 + 3.64$$
$$= 5.84 \text{ V}$$

My meter showed $E = 5.85$ V, close to the sum of the voltages across the lamps, but significantly lower than the 6.30 V battery voltage that I got under no-load conditions. Evidently, these bulbs load down the four-cell battery quite a lot. I also entertained the notion that some or all of the AA cells might have grown a little weak during the course of my experiments.

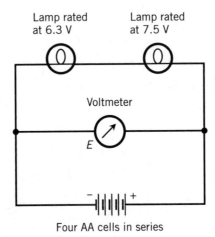

Four AA cells in series

FIG. 6-23 *Measurement of voltage E across the combination of two dissimilar lamps in series.*

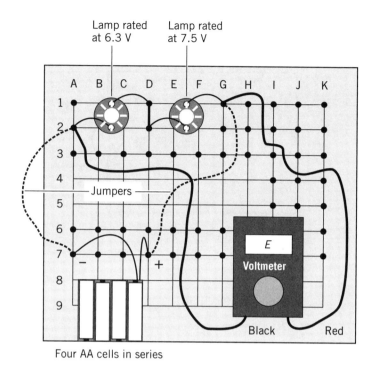

Lamp rated at 6.3 V   Lamp rated at 7.5 V

Jumpers

E

Voltmeter

Black   Red

Four AA cells in series

FIG. 6-24 *Breadboard layout for the schematic of Fig. 6-23.*

Follow the Flow

Set your meter for DC milliamperes, and connect it as shown in the schematic of Fig. 6-25 and the equivalent layout pictorial of Fig. 6-26. That connection will allow you to measure the current *I* drawn by the series combination of lamps. I got 139 mA (0.139 A). Electron current flows from the negative battery terminal through lamp N and then through lamp P, ending up at the positive battery terminal. Conventional (theoretical current) goes the opposite way, from the positive battery terminal through lamp P and then through lamp N, and finally to the negative battery terminal.

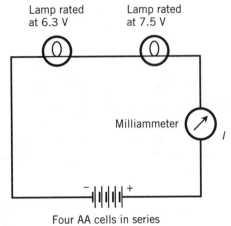

FIG. 6-25 *Measurement of current I drawn by the combination of two dissimilar lamps in series.*

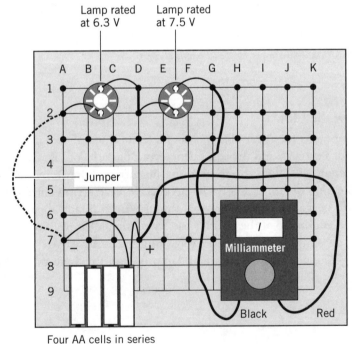

FIG. 6-26 *Breadboard layout for the schematic of Fig. 6-25.*

Now that you know the voltage across each lamp and the current going through the whole circuit, you can determine VA power numbers for the lamps individually and together. Let P_{VA1} represent the VA power consumed by lamp N, and use the formula:

$$P_{VA1} = E_1 I$$

My results came out as:

$$P_{VA1} = 2.20 \times 0.139$$
$$= 0.306 \text{ VA}$$

Letting P_{VA2} represent the VA power consumed by lamp P, the formula is:

$$P_{VA2} = E_2 I$$

In this case, I got:

$$P_{VA2} = 3.64 \times 0.139$$
$$= 0.506 \text{ VA}$$

Now suppose that P_{VA} represents the VA power consumed by both lamps operating together. In that case, theory predicts that:

$$P_{VA} = E I$$

When I input my experimental results, I got

$$P_{VA} = 5.85 \times 0.139$$
$$= 0.813 \text{ VA}$$

In theory, the VA power consumed by the lamp combination should also work out as the sum of the two VA power quantities taken individually:

$$P_{VA} = P_{VA1} + P_{VA2}$$

Adding my results of $P_{VA1} = 0.306$ and $P_{VA2} = 0.506$ gave me:

$$P_{VA} = 0.306 + 0.506$$
$$= 0.812 \text{ VA}$$

The error between my two results was a small fraction of one percent, a state of affairs that made me happy indeed!

A Compass-Based Galvanometer

In this experiment, you'll see how a current-carrying coil affects the behavior of a magnetic compass. The production of a magnetic field by an electric current is called *galvanism*. You'll need a camper's or hiker's compass calibrated in degrees, 3 feet (1 meter) of enamel-coated magnet wire, a sheet of fine-grain sandpaper, several resistors from your collection, six fresh AA cells, and some jumpers. You'll need your breadboard equipped with one holder for four AA cells and two holders for single AA cells. You'll also need a "paper punch" that can put 1/4-inch (6.4-millimeter) holes in thin cardboard.

> **Here's a Factoid!**
>
> A magnetic compass normally points toward *geomagnetic north*, which almost always differs from true or *geographic north*. This discrepancy exists because the north and south *geomagnetic poles* don't lie in the same places as the north and south *geographic poles*. The extent of the compass reading discrepancy depends on where you use it. For example, the error in Minneapolis, USA differs from the error in Paris, France.

When you place a compass near a wire that carries DC, the compass doesn't point exactly toward geomagnetic north. Instead, its needle rotates to the east or west. The rotation extent depends on how close you bring the compass to the wire, and on how much current the wire carries. The rotation direction depends on which way the current flows through the wire and on which side of the compass you place the wire. (The wire should always lie in the same plane as the compass surface.)

When you place a compass on a horizontal surface so the needle points toward N on the scale (*geomagnetic azimuth* 0°) with zero current flowing in the coil, the needle will point toward geomagnetic north provided no nearby magnetic objects interfere with the geomagnetic field near the compass. When you connect a battery to the coil, the compass needle will move. As you connect higher-voltage batteries to increase the current, the compass needle deflection angle will increase, but it will never rotate more than 90° either way. Reversing the polarity of the applied voltage will reverse the direction of the needle deflection.

Wind 8-1/2 turns of enameled (not bare) copper wire around a magnetic compass so the coil turns lie along the N-S axis of the compass as shown in

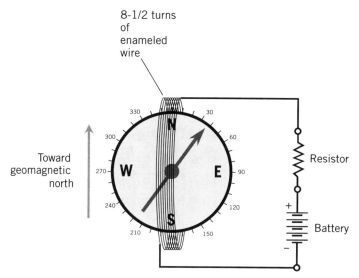

FIG. 6-27 *The galvanometer and associated circuitry. The compass must lie flat on a horizontal surface. Place it so the needle points toward N (geomagnetic azimuth 0°) when you disconnect the battery.*

Fig. 6-27. Use small-gauge, enamel-coated wire of the sort available at most hardware stores. To ensure that you get a mechanically stable coil, glue the compass onto a rectangular sheet of thin cardboard, use the "paper punch" to put holes in the cardboard just above the N and just below the S, and then wind the wire through the holes, passing the wire alternately over and under the compass. Leave roughly 4 inches (10 centimeters) of wire to spare at each end of the coil.

Extras

Note the two new single-cell holders at the lower right in Fig. 6-28. Glue them to the board and let the glue dry before you proceed further.

Use a sheet of fine sandpaper to remove 1 inch (2.54 centimeters) of enamel from each end of the wire. Place the compass onto your breadboard, and wind each sanded-off end of the wire around one of the terminal nails. If the wires are too long, trim them so they'll fit neatly between the coil and the breadboard terminals after you've sanded off the ends. Figure 6-28 shows my layout. This drawing is oriented so the right-hand side of the breadboard

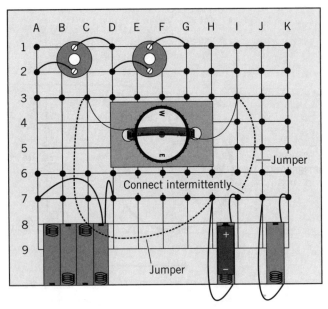

FIG. 6-28 *Suggested placement of the galvanometer on the breadboard. The jumper for the positive cell terminal should normally be disconnected. Never connect this jumper to the cell for more than a couple of seconds at a time. In this drawing, geomagnetic north lies off toward the right. Note the two additional single-cell holders in the lower right-hand part of the board.*

points toward geomagnetic north. Align the board so the compass needle points toward N on the scale. Make sure that the coil carries no current.

Use a jumper to connect one end of the coil to the negative terminal of a single AA cell. Connect another jumper to the other end of the coil. Then, for a couple of seconds, touch the "non-coil" end of that jumper to the positive cell terminal. The compass needle should rotate clockwise or counterclockwise by almost 90° so it points either slightly north of geomagnetic east or slightly north of geomagnetic west. Don't leave the galvanometer connected directly to the cell for more than a couple of seconds at a time, because the coil creates an almost perfect short circuit across the cell.

Take a resistor rated at 680 ohms, another rated at 470 ohms, another rated at 330 ohms, and five more rated at 220 ohms. Use the series combination of four AA cells from the past few experiments. With a jumper, connect the negative battery terminal to one end of the galvanometer coil. Choose two adjacent nails on the board as the location for the resistance to go in series with the galvanometer. Wrap the leads of a 680-ohm resistor around these nails. Using another jumper, connect one end of the resistor to the positive battery terminal. Switch your digital meter to a moderate DC current range. My meter has a setting for 0 to 200 mA. This worked well for me.

Follow the Flow

Firmly place one meter probe against the "non-battery" end of the resistor, and place the other meter probe against the nail where you've wound the "non-battery" end of the coil. You'll get the circuit shown in the schematic of Fig. 6-29. The compass needle should rotate toward the east. If it goes west, reverse the coil connections on your breadboard to make the current flow the other way. If your digital meter displays negative current, reverse the probes so it shows positive current. Write down the readings from the digital meter and the compass azimuth scale.

Disconnect your digital meter and replace the 680-ohm resistor with one rated at 470 ohms. Repeat the current-vs.-deflection experiment. Do the same with the 330-ohm resistor, and then with the 220-ohm resistor. Keep track of all your digital meter and galvanometer readings in tabular form.

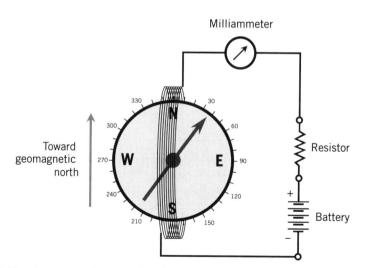

FIG. 6-29 *Arrangement for testing the galvanometer. Make sure the compass needle points exactly toward N on the scale under no-current conditions, and deflects toward the east of N when current flows through the coil.*

Wrap a second 220-ohm resistor in parallel with the existing one so you get 110 ohms. Repeat the measurements. Add a third 220-ohm resistor to the parallel combination to get approximately 73 ohms, and test the system again. Then add a fourth 220-ohm resistor in parallel, getting about 55 ohms; test again. Then add a fifth 220-ohm resistor to get about 44 ohms, and test yet another time.

Increase the battery voltage by taking advantage of the single-cell holders. Place a fresh AA cell into each holder. Wire one of the new cells in series with the four existing cells to get a five-cell battery, and repeat the experiment with 44 ohms of resistance. Then wire another new cell in series to get a six-cell battery, and once again, do the experiment with 44 ohms of resistance.

After you've made all the measurements and written down all the readings from your digital current meter and galvanometer, compile a table showing the number of AA cells in the first (leftmost) column, the rated resistor values in the second column, the current levels in the third column, and the compass needle deflection angles in the fourth (rightmost) column. Table 6-4 shows my results. Yours will doubtless differ somewhat from mine.

TABLE 6-4 *Current levels and deflection angles that I obtained with AA cells and resistors in series with a compass-based galvanometer.*

Number of AA cells in series	Resistance (ohms)	Current (milliamperes)	Deflection (degrees)
0	Infinity	0	0
4	680	9.3	7
4	470	13	12
4	330	19	19
4	220	27	28
4	110	53	40
4	73	80	50
4	55	104	54
4	44	126	60
5	44	157	64
6	44	173	66

Plot the Points!

Create a calibration graph by plotting the data from your version of Table 6-4 as points on a coordinate grid. The horizontal axis should portray the actual current in milliamperes, and the vertical scale should portray the compass needle deflection angle in azimuth degrees. Connect the points by curve fitting to obtain a continuous graph of deflection vs. current. Figure 6-30 shows the points (as small open circles) and curve that I got. Yours should look similar. You can now use your galvanometer, along with the calibration graph, as a crude analog milliammeter!

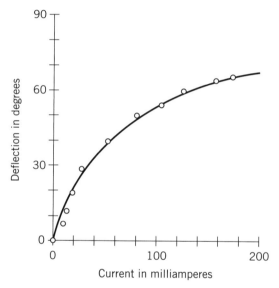

FIG. 6-30   *Compass needle deflection vs. coil current for the DC galvanometer. This graph reflects my experimental results, which appear in Table 6-4.*

Summary and Conclusion

When you want to design, build, debug, and troubleshoot electronic equipment, you'll do best if you have a good schematic (or set of schematics) to work with. Pictorial diagrams, including layouts, can help. But when you get in your lab and do the physical work, you'll never find any substitute for real-world hardware.

If you did all the experiments in this chapter, you almost certainly came up with results a little different from mine. If you had to make major parts substitutions, for example, with the lamps in the last experiment, then some of your results came out a lot different than mine. Whether you performed the experiments or merely followed along in your imagination, you got a chance to see how schematics, pictorials, graphs, and tables can work together. All these tools belong in an engineer's knowledge base.

Again, please let me make a "shameless plug" for my book *Electricity Experiments You Can Do at Home*. You'll get some hands-on lab experience and a bit of theory from that book. You'll also see some rather strange phenomena! If you want a more exhaustive presentation of electricity and electronics along with plenty of schematics and enough mathematics to keep a true nerd from getting bored, I recommend the latest edition of my book *Teach Yourself Electricity and Electronics*. Both books are published by McGraw-Hill, and you can find them at major online retailers. You might even come across them at an old-style "bricks and mortar" book store or your local public or school library!

A

Schematic Symbols

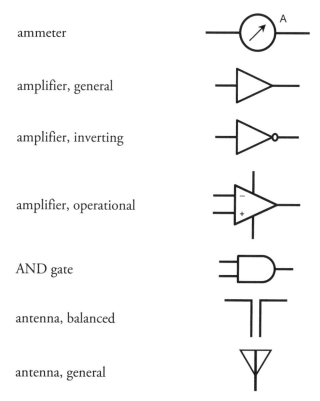

ammeter

amplifier, general

amplifier, inverting

amplifier, operational

AND gate

antenna, balanced

antenna, general

antenna, loop

antenna, loop, multiturn

battery, electrochemical

capacitor, feedthrough

capacitor, fixed

capacitor, variable

capacitor, variable,
    split-rotor

capacitor, variable,
    split-stator

cathode, electron-tube, cold

cathode, electron-tube,
    directly heated

cathode, electron-tube, indirectly heated	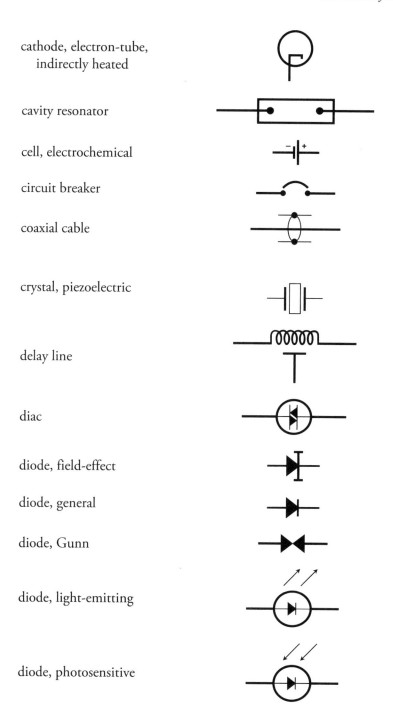
cavity resonator	
cell, electrochemical	
circuit breaker	
coaxial cable	
crystal, piezoelectric	
delay line	
diac	
diode, field-effect	
diode, general	
diode, Gunn	
diode, light-emitting	
diode, photosensitive	

diode, PIN

diode, Schottky

diode, tunnel

diode, varactor

diode, Zener

directional coupler

directional wattmeter

exclusive-OR gate

female contact, general

Ferrite bead

filament, electron-tube

fuse

galvanometer

grid, electron-tube

ground, chassis	
ground, earth	
handset	
headset, double	
headset, single	
headset, stereo	
inductor, air core	
inductor, air core, bifilar	
inductor, air core, tapped	
inductor, air core, variable	
inductor, iron core	
inductor, iron core, bifilar	

inductor, iron core, tapped	
inductor, iron core, variable	
inductor, powdered-iron core	
inductor, powdered-iron core, bifilar	
inductor, powdered-iron core, tapped	
inductor, powdered-iron core, variable	
integrated circuit, general	(Part No.)
jack, coaxial or phono	
jack, phone, 2-conductor	
jack, phone, 3-conductor	
key, telegraph	

lamp, incandescent

lamp, neon

male contact, general

meter, general

microammeter

microphone

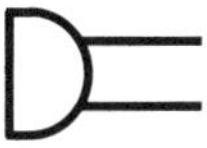

microphone, directional

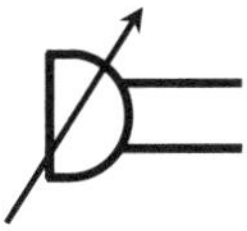

milliammeter

NAND gate

negative voltage
   connection

NOR gate

NOT gate

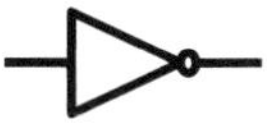

optoisolator

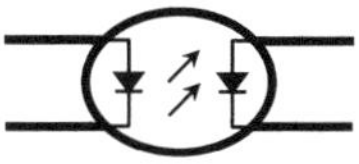

OR gate

outlet, 2-wire, nonpolarized

outlet, 2-wire, polarized

outlet, 3-wire

outlet, 234-volt

plate, electron-tube

plug, 2-wire, nonpolarized

plug, 2-wire, polarized

plug, 3-wire

plug, 234-volt

plug, coaxial or phono

plug, phone, 2-conductor	
plug, phone, 3-conductor	
positive voltage connection	+
potentiometer	
probe, radio-frequency	or
rectifier, gas-filled	
rectifier, high-vacuum	
rectifier, semiconductor	
rectifier, silicon-controlled	

relay, double-pole,
    double-throw

relay, double-pole,
    single-throw

relay, single-pole,
    double-throw

relay, single-pole,
    single-throw

resistor, fixed

resistor, preset

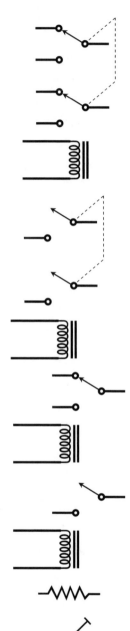

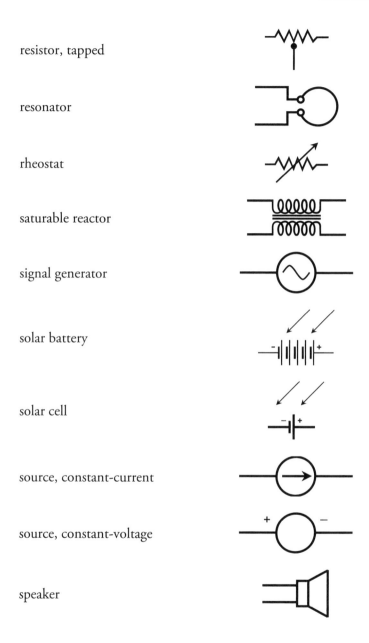

resistor, tapped

resonator

rheostat

saturable reactor

signal generator

solar battery

solar cell

source, constant-current

source, constant-voltage

speaker

switch, double-pole,
  double-throw

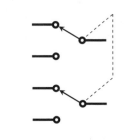

switch, double-pole,
  rotary

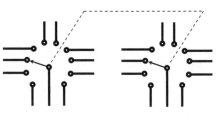

switch, double-pole,
  single-throw

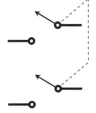

switch, momentary-contact

switch, silicon-controlled

switch, single-pole,
  double-throw

switch, single-pole, rotary

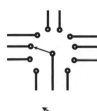

switch, single-pole, single-
  throw

terminals, general, balanced	
terminals, general, unbalanced	
test point	TP
thermocouple	or
transformer, air core	
transformer, air core, step-down	
transformer, air core, step-up	
transformer, air core, tapped primary	
transformer, air core, tapped secondary	
transformer, iron core	

transformer, iron core, step-down

transformer, iron core, step-up

transformer, iron core, tapped primary

transformer, iron core, tapped secondary

transformer, powdered-iron core

transformer, powdered-iron core, step-down

transformer, powdered-iron core, step-up

transformer, powdered-iron core, tapped primary

transformer, powdered-iron core, tapped secondary

transistor, bipolar, NPN

transistor, bipolar, PNP

transistor, field-effect,
    N-channel

transistor, field-effect,
    P-channel

transistor, MOS field-effect,
    depletion-mode, N-channel

transistor, MOS field-effect,
    depletion-mode, P-channel

Drain

transistor, MOS field-effect,
    enhancement-mode,
    N-channel

Gate

Drain

transistor, MOS field-effect,
    enhancement-mode,
    P-channel

Gate

transistor, photosensitive,
    NPN

transistor, photosensitive, PNP

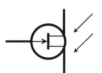

transistor, photosensitive,
    field-effect, N-channel

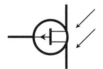

transistor, photosensitive,
    field-effect, P-channel

transistor, unijunction

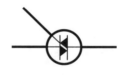

triac

tube, diode

tube, heptode

tube, hexode

tube, pentode

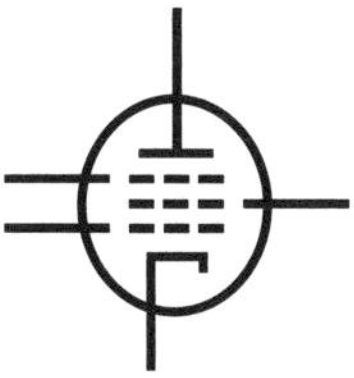

tube, photosensitive

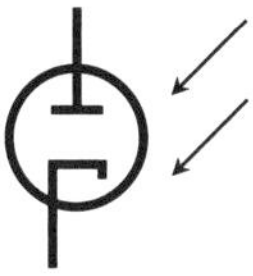

tube, tetrode

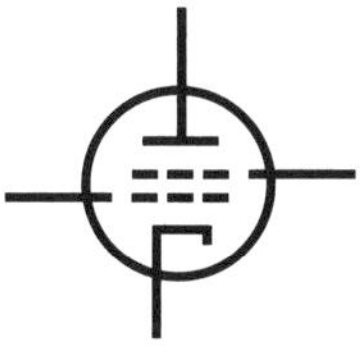

tube, triode

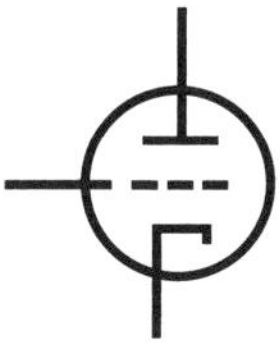

unspecified unit or
    component

voltmeter

wattmeter

waveguide, circular

waveguide, flexible

waveguide, rectangular

waveguide, twisted

wires, crossing, connected

(preferred)

or

(alternative)

(preferred)

wires, crossing, not connected

or

(alternative)

B

Resistor Color Codes

Most resistors have colored bands or regions that indicate their values and tolerances. You'll see three, four, or five bands around most carbon-composition resistors and film resistors. Other resistors have enough physical bulk to allow for printed numbers that tell you the values and tolerances directly.

On resistors with *axial leads* (wires that come straight out of both ends), the first, second, third, fourth, and fifth bands are arranged as shown in Fig. B-1. On resistors with *radial leads* (wires that come off the ends at right angles to the axis of the component body), the colored regions are arranged as shown in Fig. B-2. The first two regions represent single digits 0 through 9, and the third region represents a multiplier of 10 to some power. (For the moment, don't worry about the fourth and fifth regions.) Table B-1 indicates the numerals corresponding to various colors.

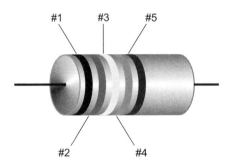

FIG. B-1 *Locations of color-code bands on a resistor with axial leads.*

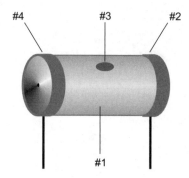

FIG. B-2 *Locations of color code designators on a resistor with radial leads.*

Suppose that you find a resistor with three bands: yellow, violet, and red, in that order. You can read as follows, from left to right, referring to the table:

- Yellow = 4
- Violet = 7
- Red = ×100

You conclude that the rated resistance equals 4700 ohms, or 4.7 k.

As another example, suppose you find a resistor with bands of blue, gray, and orange. You refer to Table B-1 and determine that:

TABLE B-1   *Color codes for the first three bands or regions that appear on most fixed resistors. See text for discussion of the fourth and fifth bands or regions.*

Color of band	Numeral (first and second bands)	Multiplier (third band)
Black	0	1
Brown	1	10
Red	2	100
Orange	3	1000 (1 k)
Yellow	4	10^4 (10 k)
Green	5	10^5 (100 k)
Blue	6	10^6 (1 M)
Violet	7	10^7 (10 M)
Gray	8	10^8 (100 M)
White	9	10^9 (1000 M or 1 G)

- Blue = 6
- Gray = 8
- Orange = ×1000

This sequence tells you that the resistor is rated at 68,000 ohms, or 68 k.

If a resistor has a fourth colored band on its surface (#4 as shown in Figs. B-1 or B-2), then that mark tells you the tolerance. A silver band indicates ±10%. A gold band indicates ±5%. If no fourth band exists, then the tolerance is ±20%.

The fifth band, if any, indicates the maximum percentage by which you should expect the resistance to change after the first 1000 hours of use. A brown band indicates a maximum change of ±1% of the rated value. A red band indicates ±0.1%. An orange band indicates ±0.01%. A yellow band indicates ±0.001%. If the resistor lacks a fifth band, it tells you that the resistor might deviate by more than ±1% of the rated value after the first 1000 hours of use.

A competent engineer or technician always tests a resistor with an ohmmeter before installing it in a circuit. If the component turns out defective or mislabeled, you can prevent potential future troubles by following this precaution. It takes only a few seconds to check a resistor's ohmic value. If you skip that simple step, built a circuit, and then discover that it won't work because of some miscreant resistor, you might have to spend hours tracking it down!

C

Parts Suppliers

All-Electronics
(800) 826-5432
www.allelectronics.com

Design Notes
(800) 957-6867
www.designnotes.com

Electronix Express
(800) 972-2225
www.elexp.com

Jameco Electronics
(800) 831-4242
www.jameco.com

Mouser Electronics
(800) 346-6873
www.mouser.com

Radio Shack
www.radioshack.com

Ramsey Electronics
(800) 446-2295
www.ramseytest.com
www.ramseyelectronics.com

Suggested Additional Reading

Frenzel, Louis E., Jr., *Electronics Explained*. Newnes/Elsevier, 2010.

Gerrish, Howard, *Electricity and Electronics*. Goodheart-Wilcox Co., 2008.

Gibilisco, Stan, *Electricity Demystified*, 2nd ed. McGraw-Hill, 2011.

Gibilisco, Stan, *Electronics Demystified*, 2nd ed. McGraw-Hill, 2011.

Gibilisco, Stan, *Ham and Shortwave Radio for the Electronics Hobbyist*. McGraw-Hill, 2014.

Gibilisco, Stan, *Teach Yourself Electricity and Electronics*, 6th ed. McGraw-Hill, 2016.

Kybett, Harry and Boysen, Earl, *Complete Electronics Self-Teaching Guide with Projects*, 4th Ed. Wiley, 2012.

Monk, Simon, *Hacking Electronics*. McGraw-Hill, 2013.

Santiago, John, *Circuit Analysis for Dummies*. For Dummies, 2013.

Schertz, Paul and Monk, Simon, *Practical Electronics for Inventors*, 3rd ed. McGraw-Hill, 2013.

Index